Handbook of Quantitative Forest Genetics

FORESTRY SCIENCES

Volume 39

The titles published in this series are listed at the end of this volume.

Handbook of Quantitative Forest Genetics

Edited by

LAUREN FINS

University of Idaho, Moscow, ID, U.S.A.

SHARON T. FRIEDMAN

USDA Forest Service, Washington, DC, U.S.A.

and

JANET V. BROTSCHOL

USDA Forest Service, Olympia, WA, U.S.A.

Production Editor

LAUREN FINS

University of Idaho, Moscow, ID, U.S.A.

KLUWER ACADEMIC PUBLISHERS

DORDRECHT / BOSTON / LONDON

ISBN 0-7923-1568-5

Published by Kluwer Academic Publishers,
P.O. Box 17, 3300 AA Dordrecht, The Netherlands.

Kluwer Academic Publishers incorporates
the publishing programmes of
D. Reidel, Martinus Nijhoff, Dr W. Junk and MTP Press.

Sold and distributed in the U.S.A. and Canada
by Kluwer Academic Publishers,
101 Philip Drive, Norwell, MA 02061, U.S.A.

In all other countries, sold and distributed
by Kluwer Academic Publishers Group,
P.O. Box 322, 3300 AH Dordrecht, The Netherlands.

Printed on acid-free paper

Printed in the Netherlands

To

David and Tracy B.,
Steve and Alex,
Ernie, Amanda and Benjamin.

Contents

Preface

This book has been four years in the making. Its inception logically dates back to the 1987 meeting of the Western Forest Genetics Association, when several of us suggested that WFGA sponsor a shortcourse that would emphasize quantitative methods. There was some discussion, but it didn't take much convincing. The members of WFGA agreed to provide both financial support and their enthusiastic professional support.

In the fall of 1987, during our first planning meeting, we realized that none of the textbooks that were currently available would meet our needs for the shortcourse. That's when we conceived the idea of compiling our instructors' lecture notes into a handbook that would serve both as a text for current and future students and as an important reference tool for professional forest geneticists and tree breeders.

In March of 1989, 10 of the most highly qualified instructors in the United States and Canada presented the week-long "Shortcourse in Quantitative Forest Genetics" to 67 participants. Although we had described our target audience as "forest geneticists, tree improvement specialists and breeding specialists holding an MS degree or a bachelor's degree with experience in forest tree improvement", our participants ran the gamut of tree improvement personnel. At the end of the week, most of them told us that the course had been very much worth their time and had provided useful information at many different levels. Even our most veteran instructors said they learned some "new tricks".

We have spent the last two-and-a-half years compiling and editing this "Handbook". Each of its chapters comes from the instructors' notes from the shortcourse. We've tried to ensure their readability and consistency by reviewing and editing all of the chapters ourselves and having them reviewed by one or several other scientists. Our reviewers were tough and thorough...our authors (bless their hearts!) were patient and cooperative.

Some of the chapters have been expanded beyond the instructors' presentations to include information that the authors would have liked to present but simply didn't have the time. Other chapters have been re-arranged to improve their flow. They follow a logical sequence and each contains information that has a special and important place in any tree improvement program.

In the first chapter by Cheryl Talbert, we begin with a justification for why one should even bother with quantitative forest genetics. Talbert *"focuses on quantitative genetics in support of six different kinds and levels of decision-making, and nicely develops the concept of a genetic-process model as providing the theoretical underpinning of quantitative genetics."*[1]

---

[1] *Editors' note: This quote and those that follow in the preface are from W.J. Libby's review of the manuscript.*

Although we assume that our readers will be well-schooled in basic genetics, we felt compelled to present clear definitions and fundamental principles to help ensure a consistent basis of understanding among our readers. In Chapter 2 then, Hans van Buijtenen concisely and elegantly reviews the fundamental principles that are the basis of all of quantitative genetics. This chapter is not intended as the student's sole introduction to the field, but rather, as "*a good text to accompany additional formal or informal instruction in genetics*".

In our third chapter, Floyd Bridgwater develops the use of alternative mating designs to achieve specific program objectives. The chapter outlines "*the main kinds of mating designs available, and compares them on the basis of backward selection (progeny testing), forward selection (gain), parameter estimation, long-term breeding, and cost*", providing the reader with the critical information necessary to decide which mating design will support various program objectives.

Once the families of interest are generated (either through controlled crosses or through open-pollinated collections) the progeny must be evaluated in well-planned, well-designed genetic tests. In Chapter 4 on field test designs, Judy Loo-Dinkins explores the various objectives of testing, the designs that can best meet those objectives and potential design flaws that may hamper one's ability to analyze data from them. According to Libby, "*Chapter 4 alone is worth the price of the book. Loo-Dinkins thoroughly covers most aspects of field-test design, layout and deployment...[providing] valuable reading for beginners and abundant food for thought for experts.*"

Chapter 5 contains a beautifully condensed version of Gary Hodge and Tim White's book on predicting breeding values. In this chapter the authors describe powerful analytical tools that can link genetic entries over time and/or space by using multiple sets of data to calculate a single breeding value for each entry. "*Hodge and White at first seem to make some simple ideas difficult, but they then use these carefully constructed approaches to make some difficult analyses possible, if not simple...This is a useful chapter, both with respect to ways of looking at traits and experiments, and with respect to ways to analyze them for the purposes of selection.*"

No book on quantitative forest genetics would be complete without a presentation and discussion of the use of analysis of variance. "*In Chapter 6, the reader is given excerpts based on 2+ decades of experience and accumulated wisdom by Roy Stonecypher. [This chapter]...gives the reader hope that big studies with missing data and bent assumptions really can be analyzed both effectively and honestly. Several examples of real analyses, some done in several alternative ways, are provided for study and evaluation.*"

In Chapter 7, Sam Foster takes the analytical process beyond the theoretical models and tackles some of the most pesky practical problems in predicting and achieving gains from tree improvement programs. He "*nicely covers the theory and practice of testing, deployment, and gain predictions...[The chapter] complements and extends the earlier chapters by focusing on how present data can be used not only to predict future growth and yield, but also on how such knowledge affects deployment of selected clones and families. This*

thoughtful chapter is backed by a substantial review of the literature, with abundant entries to primary references."

With changing technologies and changing markets, tree breeders are often required to present economic justifications for continuing existing programs and/or developing new or expanded programs. In Chapter 8, Sharon Friedman nicely presents a variety of economic tools for decision-making in breeding programs. This chapter *"leads most forest geneticists and many tree improvers into new topics of proper concern. For some, it provides the framework for thinking about such topics. For those who actually need to do such analyses, it provides theory, examples, and appropriate references to the relevant literature...[The] methods of calculating the annual planting area needed to justify an improvement program, and the contrasts of methods...[were] particularly useful and insightful."*

In our last and probably most complex chapter, Bob Westfall presents some sophisticated and elegant analytical tools for developing seed transfer guidelines. This chapter logically closes our book in that it explores in depth some of issues introduced in van Buijtenen's chapter on fundamentals, namely the determination of seed procurement zones for the first generation and seed deployment zones for advanced generation programs. Much of the focus of Westfall's chapter is on dealing with genotype by environment interactions and the issues surrounding genotypic stability across environments. *"For the beginner, Westfall raises a number of important questions and suggests good ways to go about answering them. For more advanced readers, he provides useful insights and details with respect to some complex techniques. It is not a chapter where novices can find cookbook routes to advanced analyses, however. His caution about not allowing computer analyses and multivariate techniques to distance the investigator from the data is well taken."*

That's it...nine chapters of excellent information and advice from experts in the field. New and more efficient analytical tools continue to be developed. As they come to light, new workshops and shortcourses will be offered and new books will supplant the old ones. In the meantime, we enjoyed working on this book and the shortcourse that preceded it. They were both challenges for us and took more time than we'd originally anticipated. But we expanded our own professional knowledge and expertise in the process. Our objective now is to help others do the same.

Lauren Fins
Sharon Friedman
Janet Brotschol
June 1991

Acknowledgements

The editors acknowledge and thank the members of the Western Forest Genetics Association for providing their professional and financial support for the original shortcourse on which this book is based. We also thank our chapter authors for their patience and professionalism in writing and revising their manuscripts and responding to questions and critiques from us and reviewers. They did a magnificent job of condensing, into single chapters, bodies of information that might each constitute an entire book. Our chapter reviewers, in alphabetical order, were: Floyd Bridgwater, Bob Campbell, Jim Cathcart, Chris Hansen, Steve Knowe, Steen Magnussen, Barb McCutchan, Steve McKeand, Gene Namkoong, and Gerald Tuskan. Their individual and collective contributions to the readability and flow of the chapters were invaluable, and we thank them heartily. Our sincere appreciation and thank yous go to W.J. Libby who reviewed the entire manuscript. His watchful eye caught several remaining inconsistencies within and between chapters, and his insightful comments helped us to maintain appropriate perspective throughout the book. And finally, a giant thank you to Deb Hornbuckle who helped pull all the pieces together. Without her knowledge of word processing and her eye for detail, we could not have produced this book.

The Editors
June 1991

Chapter 1
Quantitative Genetics: Why Bother?

C. B. Talbert[1]

Foresters and tree improvers alike often believe that genetic manipulation is a new phenomenon in forestry, requiring major changes in philosophy and practice. I suggest instead that foresters have always been geneticists! But, to avoid an identity crisis, or a full-scale war, let me elaborate.

1.1 Everyone's a Geneticist!

From the time that the first 'forester' gathered tree seeds and planted them, or cleared vegetation to favor certain trees over other plants, genetic control has been exercised to shape stands for sites and uses which best suited those early forest managers. Through time, experience with seed sources has led some of those sources to be preferred over others — another form of genetic control. In some countries, such as New Zealand, Australia and parts of Europe, non-local species and seed sources form a significant component of the local forests, representing a major alteration of the genetic composition of those forests. At the most extreme, for five centuries and

[1]*Quantitative Geneticist, Weyerhaeuser Company, Strategic Biological Sciences Research, Tacoma, Washington*

L. Fins et al. (eds.), Handbook of Quantitative Forest Genetics, 1–28.

more, foresters in some areas of Japan have been planting stands from rooted cuttings of a small number of cultivars, derived from individual selected trees preferred by the foresters in the region.

The message here is that the practice of genetic manipulation in forestry is neither new, nor conceptually startling, nor fundamentally separate from the basics of forest management practice. Advances in tree improvement and forest genetics have simply made it possible to make genetic improvement happen more quickly, efficiently and effectively.

1.2 Decision-Making in Today's Environment

The environment in commercial forestry today is one of extreme stress, uncertainty, competition and change. It is also one of great excitement, technological innovation and global commercial opportunity. Logs and wood products now move freely among continents and hemispheres. Customers worldwide are demanding value and quality at a reasonable price. New materials, new manufacturing technology, computing and information technology, and biotechnology are reshaping basic assumptions and options for the future in forestry. The success of individuals and organizations in this environment hinges on their ability to focus and capitalize on the excitement and opportunity — to make decisions and take action — without becoming bogged down by the uncertainty and stress.

The focus is on the decision-maker...and who is this mysterious person with such great power? In today's environment, with computing power and information at each person's fingertips, and growing support for grass-roots decision-making, to a large extent the decision-maker is ***you***: the person who must take action and get a job done. And, if there is one universal success factor in forestry or any other enterprise today, it is the ability to take effective action, to make the right things happen efficiently and quickly.

Taking action, in the right direction, quickly and cost-effectively, requires two things. First, information: information on the goal, information on available alternatives to get to the goal, and information by which to compare the alternatives and track progress toward the goal. Sometimes this information is quantitative, tangible, easy to interpret. Often, however, the information is qualitative, intangible, interpretable only through experience and intuition. Second, we must have the capability to process the information to provide us with direction.

In forestry we are typically chronically short of information with which to evaluate the impact of a given action. For one thing, the longevity of the crop creates an inevitable technology lag. Many empirical trials of sufficient duration to be meaningful, by their very longevity, have been established using technologies that are obsolete, or to address questions that are no longer of interest. Couple this with a very high cost, in dollars, land, effort and time, to establish, maintain and measure well-designed long-term trials in forestry, build in very high levels of environmental and biological complexity in forest systems, and the result is information starvation.

In forest genetics we face yet another level of complexity. With current technology we cannot directly measure the genetic value of a tree in a plantation free of the confounding influences of the environment in which it is growing. This makes it very difficult to project the impact of our efforts at genetic control. We do not have a clear picture of the underlying machinery. Yet without this picture we run a serious risk that our decisions will be based on incorrect assumptions about their outcome, and may therefore mislead us, and/or cause us to waste precious resources.

The solution lies in the development, testing and continual refinement of models explaining the behavior of the systems upon which we depend. By the word 'model' I do not necessarily mean to imply some enormous and complex biomathematical entity, but rather, a description of the mechanisms that produce a set of behaviors of interest to us, based on the best thinking, logic and data available. Once we understand the mechanisms behind response to activities or treatments of interest, it becomes possible to project the response to new treatment combinations.

Such an approach is also likely to reduce the total cost involved in forest management decision-making in two ways. First, the articulation of a set of hypotheses about the workings of a process will make it possible to break down a highly complex problem into a number of simpler issues that may be addressed by small, focused experiments. Second, once tested and confirmed, a process model can allow prediction of future scenarios without the need for costly empirical trials to examine each alternative of interest.

1.3 The Origin and Basis of Quantitative Genetics

In genetics, we have the benefit of decades of model development behind us, by pioneers from Gregor Mendel and Charles Darwin, through Sir Ronald Fisher and Sewall Wright, and on to Oscar Kempthorne and C. Clark Cockerham, to recognize only a few. These people, and scores of others working with organisms from *Drosophila melanogaster* to *Zea mays* to *Homo sapiens*, have observed the behavior of genetic populations and put forward ideas and conceptual and mathematical models of the mechanisms controlling that behavior. Mendel's simple models of inheritance of flower color and of fruit morphology gave way to multi-gene models with many alleles, interactions among loci, and environmental influences on trait expression. And these multi-gene models are in turn being modified as new information becomes available from molecular genetics, biochemistry and physiology research.

It is this desire to understand genetic processes and make decisions to make productive use of those processes that has led to the field of quantitative genetics. According to Kempthorne (1988):

"The aim of quantitative genetics is then: to develop validated models for phenotypic expression in the face of partial nonidentifiability of genotypes and partial non-identifiability of environments, ...to develop a validated model of the dynamics of populations under natural or artificial selection, and to use such a validated model to make choices among the huge number of artificial selection procedures that the human mind can invent."

We depend heavily on genetic models in everyday tree improvement decisions. The use of a family mean to weight an individual's phenotype for selection depends on models of inheritance that allow us to predict the genetic similarity of relatives, and how that similarity will be expressed. The decision to separate sibs in a seed orchard depends on a genetic model relating matings among relatives to increases in frequency of deleterious genes in homozygous combinations, a model that predicts negative impacts on performance. The appropriate number of families to allocate to a plantation block depends on a model of the genetic makeup of a family, which allows us to predict the likelihood that a given family will contain deleterious genes, or that it will exhibit a reduced amount of genetic variability and buffering against environmental change.

And yet we seek always to simplify, to develop standard rules and procedures by which tree improvement decisions can be made. In simplicity there is great benefit. This quantitative genetics handbook transforms years of experience and advanced training into methods that can be followed even by practitioners with little training in quantitative genetics.

However, as we simplify, we also incur a considerable danger, well described by Feldman and Lewontin (1975):

"...Relations between genotype, environment and phenotype are at base mechanical questions of enzyme activity, protein synthesis, developmental movements and paths of nerve conduction. We wish, both for the sake of understanding and prediction, to draw up the blueprints of this machinery and make tables of its operating characteristics with different inputs and different milieus. For these problems, statistical descriptions,

especially one-dimensional descriptions like heritability, can only be poor, and, worse, misleading substitutes for pictures of the machinery. There is a vast loss of information in going from a complex machine to a few descriptive parameters...It is rather like trying to infer the structure of a clock by listening to it tick and watching the hands."[2]

Breeders are often enthralled with point-in-time manifestations of genetic processes — numbers like heritabilities, correlations, breeding values and gain — because they offer glimpses into the genetic machinery. These parameters are fairly easy to estimate, using procedures that can be taken from a cookbook, and they are increasingly available from the myriad of genetically structured field trials growing around the world.

There is most definitely great value to this growing bank of information. Parameters like heritability and correlations can be useful and appropriate indicators of genetic behavior when they are estimated precisely and accurately, applied in the context of the time and space in which they were derived, and treated with the necessary humility. They may be used to guide selection decisions, to estimate gain tradeoffs among alternatives, and to guide design of orchards and breeding programs. Trends in genetic parameter estimates with time, or relationships of genetic parameter estimates with environmental factors, can help us develop hypotheses about genetic processes. However, genetic parameter estimates are as much or more a reflection of our experimental design and our own assumptions about the mode of operation of genetic processes, as they are an actual

---

[2]*Dr. Claire Williams is gratefully acknowledged for providing this quote.*

indication of how genetic processes operate. In fact, as Feldman and Lewontin point out, if we don't understand how genetic processes actually work, today's heritability values may actually mislead us in terms of where the same processes will take us tomorrow. These concerns are only compounded when genetic parameters are estimated with a small number of families, resulting in a high risk that even today's heritability estimates may be far from correct. The message is: use genetic parameter estimates to move forward, but use them as guidelines rather than laws of nature. Appreciate the uncertainty associated with them. Use them to help you clarify your thinking about the underlying operation of the genetic system you are working with. And estimate them properly, or don't estimate them at all.

1.4 Making Decisions Based on Quantitative Genetics

As I've alluded to above, most decisions that are made in implementing a tree improvement program depend on quantitative genetic models, although often these models have been in use for such a long time and are so well accepted that breeders sometimes forget that they are invoking them. Tree improvement decisions fall into 6 major categories:

* Program initiation/continuation decisions
* Selection decisions
* Breeding decisions
* Testing decisions
* Production decisions
* Allocation/deployment decisions

In the section below, I discuss the types of decisions that are often required in each of these categories, and the ways in that quantitative genetics can guide these decisions.

1.4.1 Program initiation/continuation decisions

Many if not most tree improvement programs are asked, at the time they are initiated and periodically thereafter, to justify the resources committed to them. Some do this by predicting a biological gain from genetic improvement, and qualitatively assessing whether this gain is worth the costs and effort. Others attempt to quantify expected dollar value gains and compare them to costs in a financial analysis framework. The basic element here is the estimate of gain — in other words, the change in average of one or more traits for a population resulting from a set of genetic improvement activities. Gain also plays a key role in other types of decisions, so we will discuss it in some detail here.

Gain is probably the most problematic of all quantitative genetic statistics in forestry, and arguably, among the most complex. Gain results from changes in frequency of alleles, acting singly or in combination to control a characteristic of interest. These allele frequency changes occur as the result of treatments that modify the genetic structure of the population. And each step in a tree improvement program, from selection and controlled mating to orchard establishment and roguing, is a treatment that impacts genetic structure.

However, empirical measurement of genetic improvement responses is a challenging exercise. For one thing, our treatments are often unclearly or subjectively defined, and often vary both in time and space. We can't actually measure (yet) the changes in allele frequencies associated with our actions. In addition, to separate environmental effects from genetic treatment effects we must grow *treated* and *untreated* populations in the same environments. Also, we most often wish to measure response in terms of some harvest-age traits, and also usually on a unit area basis, and such trials are extremely expensive to conduct with precision. For these reasons, the relationships — the dose-response curves — between genetic treatments, like selection, and the frequency of desirable alleles are not well-described by data.

There are a number of alternatives for estimating gain. Differences between select groups and unimproved checks can be measured in tests, and converted to gain via an estimate of *heritability* or a *coefficient of genetic prediction*. *Breeding values* may be predicted for parents based on progeny test data via best linear prediction, and the average breeding value of the selected parents can be used as a predictor of gain. Unit-area gain at harvest may be predicted from young stands using growth and yield models, or by the application of correlations between young, individual tree traits and harvest-age unit-area traits. Certainly, other methods, which I have not listed here, have been applied, but these are some of the most common. Specifics on these methods are described in the chapters that follow in this handbook.

An estimate of biological gain may be all you need, if your organization bases initiation or continuation of tree improvement on some desired biological gain. Even if you are in this relatively simple situation, it is critical to recognize the often-significant error associated with gain estimates, and ensure that you would make the same decision even if the true gain were at some lower level within the bounds of estimation error.

More likely, you will need to carry out some level of financial analysis to move from a biological gain to a quantitative comparison of benefits and costs. It is beyond the scope of this discussion to review all the financial criteria that are available to compare benefits and costs; however, each has similar analytical requirements. First, you must convert your biological gain to value, via stumpage prices, product recovery and value curves, or some other similar method. Second, you must understand the incremental costs associated with conducting the proposed program, or continuing in a program, and when these costs will be incurred. Finally, you must have a way of discounting benefits and costs that come at different points in time, so that they may be compared in a single number. The analytical methods you would apply for genetic improvement are the same as those you would use to evaluate silvicultural treatment benefits financially. Again, it is critical that you account for the impact of uncertainty on your decision: uncertainty in gain estimates, uncertainty in costs, in your value function, in the timing of benefits. Under highly uncertain conditions, often a sensitivity analysis approach is an appropriate technique. I'll cover this subject in a bit more detail later.

1.4.2 Selection decisions

Selection decisions in tree improvement can encompass several levels. The first level, of course, is the species. Typically, species selection decisions are made on the basis of an organization's geography, goals, and sometimes, the results of empirical species trials. This level of selection will not be discussed further here.

A second level of selection involves choice of the geographic source (provenance) of the germplasm to be planted. Source selection issues are receiving considerable attention today, and have for decades around the world. In some cases, the task at hand is to choose a source or set of sources that perform well in a specific geographic area. In such cases, the quantitative process employed is sometimes as simple as comparing samples of sources in an empirical trial on representative sites, and selecting the source that performs the best in terms of a set of predetermined criteria, often including survival and other indicators of adaptability. It is also possible to generate descriptive statistics about sources, and families within sources, which indicate their stability, or tendency to change rank, across an array of geographies.

In other situations, it is of interest to develop general guidelines for movement of sources across a broad geographic area. Factor analysis and principal components analysis procedures have been used to group genetic material of known geographic origin, based on their similar performance when planted in specific locales. Maps have then been drawn or distance and elevational transfer guidelines have been developed to outline recommended limits for movement of material.

A third level of selection involves the choice of individuals or families within a source. A key to effective selection at this level lies with the ability to compare candidates on an equal footing, as free as possible from the confounding effects of different environments, ages, test precision, competitive situations, and the like. Quantitative approaches to selection are superior to subjective approaches in that they can be repeated, the assumptions used are considered and made explicit, and the results of selection can be related back to the selection treatment in a quantitative way. Several quantitative selection methods exist, and they differ from one another in ways that may be insignificant under some circumstances and major under others.

In forest stands of unknown genetic structure, candidates are typically evaluated by comparing their characteristics with only those neighbors who are judged to be growing in similar environments, under similar conditions. The similarity can be judged subjectively or quantitatively.

As we move into progeny tests and genetic tests designed for advanced generation selection, it becomes possible to learn about the genetic value of an individual or family through the performance of its relatives. Since the relatives (typically half- or full-siblings and sometimes combinations of these and other relatives in open-pollinated families) of candidates can be replicated across a range of environments, it is possible to compute an average value of the genes shared by those relatives, which is far less clouded by the confounding effects of environment than the value for any single individual. One role of quantitative selection methods is to provide appropriate weighting to be given to the relatives' performance in evaluating a given candidate.

In addition, it is often the case that we wish to make improvements in more than one trait through our selection efforts. Different traits are more or less confounded by environmental effects, may be more or less important to final value, and may be intercorrelated so that selection on one trait will have a positive or negative impact on another trait. A second role of quantitative selection methods is to provide weighting for different traits of interest.

Sometimes we would like to compare candidates that are growing in very different tests, either in terms of precision, or mating design, or age, or site, or candidates that have differing numbers of relatives under test. Selection methods have been developed that enable the analyst to adjust for such differences and quantitatively score candidates from different tests on an equal footing.

So far we've discussed selection decisions involving which candidates to select. But, before these decisions can be made, one must determine which traits to improve through selection. This decision depends on the goals of the organization. In most operations these goal traits are those that relate most closely to value, according to the organization's own value equation. Some place high value on adaptiveness, others on quality, others on yield. Economic analysis may be required to determine whether a certain trait is economically important enough to warrant the cost of genetic improvement. Another, related decision involves which traits should be measured and used for selection — the selection criteria. Sometimes the selection criteria are the same as the goal traits. In other cases, the goal traits are costly and/or difficult to measure, and less costly or more easily evaluated selection criteria, correlated with the goal traits, are screened instead. An example is Pilodyn penetration, used as a selection criterion for improvement of specific gravity, a goal trait. Quantitative analysis may be applied to evaluate an array of selection criteria for their expected effectiveness in improving the goal traits. Based on the heritability of each selection criterion, the amount of variation available for selection, and its correlation with the goal traits, a gain estimate can be produced for each selection criterion or set of criteria. These gain figures may then be combined with cost estimates for evaluating each selection criterion in a cost-benefit analysis.

In addition, the tests in which the candidates are to be grown for comparison must be designed and established. Test designs must be specified which provide effective screening for the traits of interest at the desired level of precision. Quantitative methods exist for estimating the number of trees and replicates required to be 95 percent confident of detecting real differences among groups.

The final decision I'll discuss in this section relates to the number of candidates to select, and the genetic structure of the selected population. In some cases, the desired number of selections is pre-set, say, by an orchard design. In other cases, for example, for an ongoing breeding program, the number may be set by the breeder to accomplish certain goals. As the number of selections comprising a breeding population gets smaller, the greater the selection intensity and potential gain from that selection step. However, there are tradeoffs, in reduced genetic variation, in an increased probability that rare, favorable genes will be lost, and in a more rapid buildup of relatedness and inbreeding as the smaller population is cycled through time. Quantitative genetic models have been developed that enable the breeder to estimate the impact of each of these tradeoffs, so that various population size alternatives may be compared. In a related context, genetic structure alternatives — i.e. the degree of constraint on the number of relatives that may be included in the selected group — may be evaluated. Since the best candidates often belong to just a few families, there is an apparent immediate gain benefit to including many siblings in the selected group. However, the result would be an increasing inability to avoid related matings and buildup of inbreeding. Quantitative geneticists are beginning to quantify the relationship between inbreeding level and value for important traits, allowing this tradeoff to be evaluated as well.

1.4.3 Breeding decisions

Breeding decisions can take a number of forms. Design of a breeding pattern or strategy is a key element of an ongoing improvement program, since it determines how the desirable alleles in the selected parents will be recombined. The concept of assortative mating, or crossing of like individuals, has been around for a long time, but has only recently been discussed with reference to many large-scale tree improvement programs. Comparison of assortative mating schemes with random mating schemes may be made in terms of changes in genetic variances, effective population size and inbreeding level over time, as well as in terms of projected gain. Whether assortatively or randomly combined, the specific pattern by which parents are assigned to one another (e.g. in a diallel, a factorial, a nested design, a wind-pollinated or polymix system, or some adaptation of these) has significance in the number of unrelated families available for advanced generation selection, as well as the number of crosses (and dollars) required to evaluate a parent, the ability to track pedigree, and the ability to estimate genetic parameters. The one you choose should be based on the priority your organization places on these different design features.

Breeding decisions may also take the form of crossing schemes for large-scale production of full-sib families, say, via Supplemental Mass Pollination. The degree to which crosses among high-value relatives are allowed for production purposes depends on the expected impact of related matings on seed set, germination and long-term productivity.

1.4.4 Testing decisions

Testing decisions share considerable relationship to selection decisions, since a majority of the testing that goes on in genetic improvement programs is for the purpose of providing comparisons among various genetic entries, like sources, families or clones. Primarily, testing decisions involve test design.

Once the goals of the testing program have been established, it is important to define a target precision of the critical treatment comparisons to be made from the test. For example, it may be your target to be able to demonstrate a 10 percent volume difference to be statistically significant at the 95 percent level. Given some ballpark estimates of the site-to-site, rep-to-rep, and within plot variances, it is possible to come up with design specifications that are likely to provide the target discrimination from the test. This is a critical step; given typical testing costs of more than five dollars per tree through 8 to 10 years of testing, the cost of a careful test design planning exercise is quite trivial in comparison, but the difference may mean a useful test vs. one that cannot demonstrate significant differences.

1.4.5 Production decisions

In the category of production decisions, I place decisions about the manner by which the genetic material generated through selection, breeding and testing is packaged and multiplied into propagules for field planting.

The first production decision that must be made is the type of propagule to produce: that is, seed or some type of vegetative propagule. In some cases, this decision is moot because only one propagule type is available. In other circumstances, both technologies are available, each with its own cost structure, risks and associated gains. A related concern is which genetic material to propagate. In a seed-based system, wind-pollinated, polycross, or full-sib families may be generated (control-pollinated families involving some incremental effort and cost, along with some incremental gain, over a traditional wind-pollinated seed orchard). When vegetative propagation is available, a wider range of choices can exist; in addition to multiplication of valuable families, vegetative propagation technology can sometimes enable the mass-production of individual genotypes. Such true cloning, in theory, offers the highest possible genetic gain within a single breeding cycle. However, for many species, the costs of propagation are also higher for vegetative methods. In addition, without clonal testing, for many traits of low or moderate heritability there is a significant risk that the individuals selected for cloning will not be as outstanding as expected. Are the benefits worth the costs and risks? This answer can only be derived through estimation of expected gains and costs from the different systems, and comparison of those costs via some kind of financial criterion. When risks are involved, the probability of each possible outcome can be estimated along with the financial impact of that outcome to provide some guidance.

Let us assume that a seed orchard route is the route you've chosen. Seed orchard design involves some specific decisions, about the number of individuals and relatives to allow, the number of ramets per clone if your orchard is clonal, and the minimum distance that may be allowed among different kinds of relatives in a wind-pollinated orchard. Typically, orchard sizing (the number of individuals and the number of ramets) is based on the expected seed production per tree at some point in the future, balanced against the expected seed needs and your organization's desired margin of protection against unforeseen drops in production. The number of relatives allowed in an orchard, and how relatives are located in the orchard, depends on assumptions about pollen flight and the expected negative impact of matings among relatives.

In both seed and vegetative propagation programs, the numbers of families or individual genotypes to propagate also involves implicit or explicit assumptions about the risks and benefits of given levels of genetic diversity in the production population. It is a generally accepted concept that, as the population size and number of unrelated individuals declines, the genetic variability in the population also declines: certain genes or gene combinations become more common, while other genes or gene combinations drop in frequency. If only a small number of genotypes is represented in the deployed population, there is a significant chance that certain rare alleles may not be included at all, simply due to random sampling. The point at which that reduction in variability is likely to have important ecological or economic consequences depends on several unknowns: what environmental pressures are likely to face the population over time, what new economic demands are likely to be made on the population in the future, what genes or gene combinations will influence the population's ability to respond well to those environmental pressures, and how those genes are controlled. This is an area in which much new information and guidelines are needed. Molecular genetics is likely to provide important insights. In the meantime, some theoretical work has been done which estimates the number of genotypes that should be included in a population to provide a given level of probability of including a rare allele. This offers some guidance, as can sensitivity analysis of the probability of future catastrophic events, their relationships to allele frequency, and the economic impacts of those events should they occur. Also, as specific families and genotypes are grown for many years on many environments in genetic tests and operational stands, we will gain more and more confidence in their ability to withstand whatever nature can put forward.

1.4.6 Allocation, deployment and management decisions

How should the genetic material generated and packaged in your tree improvement program be deployed in the field? Should it be planted in single-genotype or single-family blocks, or should families or genotypes be randomly or systematically mixed? How should genetic material be allocated to sites? What forest management regimes are most appropriate for improved stock? I place these questions in the category of allocation, deployment and management decisions.

The block vs. mixture question relates in part to the level of environmental risk conferred by having similar (in the case of families) or identical genotypes (in the case of cloning) planted in contiguous blocks. To grapple with this aspect of the question we must again speak in terms of probabilities: what are the likely environmental pressures that will be faced by the stands through a rotation, and to what extent will genetic similarity (or perhaps the absence of some important, rare allele in a block) impact the stand's ability to respond to the environmental pressures without significant loss? In addition, the block-mix question relates to the benefits associated with growing families or genotypes in blocks or mixtures. Benefits of block or mixed culture depend in part on the differential competitive behavior of individual families or genotypes. Empirical trials are only beginning to provide the necessary information to make this judgment. In the meantime, some organizations have chosen to manage families in blocks because of resulting efficiencies in the orchard and nursery, and due to the greater ability to tailor management and utilization strategies to more predictable and homogeneous stands.

The site-allocation question relates directly back to our previous discussion of source movement. Are certain families or genotypes more stable than others? Can greater gain be expected from allocation of the best material to the best sites, or vice versa? Genetic testing on a broad range of sites enables quantitative analysis of stability and genotype-site interaction. Once this is quantified, each organization must examine its willingness to develop complex allocation procedures that allow matching of specific material to specific site types.

The development of forest management prescriptions for improved stock will depend on the expected stand growth and yield behavior of that stock. If improved stock simply proceeds more rapidly through the same stand development patterns as unimproved stock, then the analytical procedures we use to set management prescriptions for our unimproved plantations should produce sound recommendations for improved stock. However, if certain families exhibit different competitive responses, or allocate wood differently to diameter and height, or show more or less efficient utilization of site resources for growth, then it may be necessary to adjust our standard growth and yield models for our most important genetic material. Empirical trials are critical for development and testing of growth and yield hypotheses about families and clones, and many of these are underway.

1.5 The Care and Feeding of Data

I mentioned previously some of the hazards of straightforward, cookbook approaches to arriving at key program decisions. At the bottom of all of these hazards is one issue: Bad data can often provide answers via the same methods as good data.

What are *bad data* and *good data*? Good data describe the attribute you intended to describe. Bad data may describe some other attribute, although you *think* it is the intended one; or, the data collection process may be unclear enough or poorly documented enough that you're not certain just what attribute you *are* describing (or, perhaps, what environment). Good data come from a sufficient sample, such that you can reliably expect to get a similar answer if you were to sample again under the same conditions. Bad data are unreliable, often because a sample was too small. Good data enable treatment (say, family) differences to be estimated with sufficient precision that you can distinguish among them with confidence. Bad data may contain so much 'noise' that even apparently large treatment effects cannot be distinguished from random error effects.

These contrasts are made in black and white to make a point. In fact, there is a continuum between bad and good. Only rarely does a set of data turn out perfectly appropriate, accurate and precise relative to the application for which it was collected, and data are also only infrequently absolutely irrelevant and useless for their intended purpose. However, there is a tendency for any and all data to be called into use, regardless of quality for the intended application. SAS will generate variance and covariance components, and heritabilities can be calculated, from a sample of 2 families with the same speed and ease that it will with 200 families (more easily, in fact!). But what do the two parameter estimates mean? How should they be applied?

The bottom line is this: each person who finds him or herself in the position of data analyst and decision-maker must take the personal responsibility to ensure that their data are as good as they can be, and that they are used appropriately. If a set of data contains a trait that is poorly described, or is associated with abnormally high measurement error, the trait should be checked, and remeasured if necessary. If there are only a small number of families in a set of data, the risk is high that estimates of genetic variances and covariances from those data will be either highly imprecise, inaccurate, or both, and the data are probably inappropriate for use in such an analysis. If a test produces measurements that are consistently so 'noisy' that even large treatment effects cannot be judged statistically significant, the data should not be used for statistical testing of treatment effects, and if possible, the experimental design should be altered before any new tests are established.

But what do we do if these bad data are the only data we have? If your data describe a trait or a treatment inaccurately, either because of poor data-collection and documentation, or because of insufficient sample size, you may be better served by calling on indirect information (do others have tests that could shed light on my question?) and by using your own experience and judgment, than you would be by using the potentially misleading answer provided by your data. In such a case, or if your data are merely imprecise, a sensitivity analysis approach may be the most appropriate. Sensitivity analysis is only a technical term for picking a likely range of values for your answer, and figuring out what your decision would be across that range. If your decision wouldn't change across the range of likely answers, then the quality of the answer doesn't matter. If the decision changes across the range of likely answers, then you have a choice: you can collect new data that do a better job, or you can make the decision based on your experience and judgment.

One further point on the subject of good and bad data. It is a good idea for any person with a significant data analysis role to develop some standard practices for ensuring high quality data. Get in the habit of describing the traits to be measured in writing and in detail both before and after the data-collection trip — and tell your data-gathering crew what your expectations are in terms of measurement error. State clearly what you want to learn, before you lay out the test or go out to collect data. Apply sound experimental design procedures that are focused on your objectives. This may include some *a priori* subsampling to define the ultimate sample size you'll need to accomplish your objectives. Once in, check your data thoroughly for errors or inconsistencies right away; you may want to return to the source to check a questionable measurement. Back up your data so that it won't be lost. And write down your standard practices, so that you can repeat what works and improve what doesn't.

1.6 Where Quantitative Approaches Leave Off

In several places during the above discussion I've noted that clear, quantitative guidelines may not be available to assist you with a complex decision. Even when quantitative tools exist to address a question, your data and/or uncertainties associated with the method may create considerable uncertainty around which is the best course of action.

There is no substitute for intuition, experience and observation in complex decision-making. It is said that some men use statistics as a drunk uses a lamppost: for support rather than illumination. I've taken this as an important piece of advice. Any number derived via methods described in this workbook should only be used as a guideline, as one piece of information in your bag of tricks. If your instincts or observations cause you to suspect your numbers, you would be well-advised to go back and check your numbers.

I've also alluded to the important role played by sensitivity analysis in many decisions that are accompanied by a high degree of uncertainty. When the range of error in a key statistic is high, it is critical to examine your decision against levels of the statistic at the high and low end of the range. If your decision doesn't change across your range of error, then all the better. If your decision would be different across that range, you may need to back up, re-estimate your statistic with a smaller range of error, or make your decision using your best judgment.

This discussion has been intended to pique your interest and indicate the ways in which quantitative analysis plays a role in key tree improvement and forest management decisions. The following chapters provide some for the tools and procedures to assist you in making those decisions.

Literature Cited

Feldman, M. W. and R. C. Lewontin. 1975. The heritability hang-up. *Science* 190(4220):1163-1168.

Kempthorne, O. 1988. An overview of the field of quantitative genetics. Pp. 47-56 in Weir, B. S., E. J. Eisen, M. M. Goodman, and G. Namkoong, eds. *Proceedings of the Second International Conference on Quantitative Genetics.* Sinauer Associates, Inc., Sunderland, Massachusetts. 724 p.

Chapter 2
Fundamental Genetic Principles

J. P. van Buijtenen[1]

The purpose of this chapter is to define the basic terminology of genetics and quantitative genetics in preparation for the chapters that are to follow. This chapter is divided into three sections: Mendelian genetics, population genetics and quantitative genetics.

2.1 Mendelian Genetics

In this section we will outline the definitions of terms used in basic genetics and discuss the genetic patterns described by Mendel's laws of inheritance.

2.1.1 The gene

The fundamental unit of Mendelian genetics is the gene. For Mendel the gene was an abstract concept, a unit of inheritance passed on as a discrete package. In the meantime we have discovered that biologically, it corresponds to a sequence of DNA with a very complex structure which we are still learning about.

[1] *Head, Reforestation Department, Texas Forest Service and Professor, Forest Science Department, Texas Agricultural Experiment Station, Texas A&M University, College Station, Texas.*

L. Fins et al. (eds.), Handbook of Quantitative Forest Genetics, 29–68.

2.1.2 Locus

Genes are organized in chromosomes and each gene is located at a specific position on a chromosome. This position is called a locus. When the gene occurs in different forms due to mutations, they are called alleles. These, in other words, are slightly different versions of the same gene located at the same position (locus) on the chromosome.

2.1.3 Ploidy

The set of chromosomes may occur in different numbers of copies in different species of the same genus or in different tissues of a particular organism. This is called the ploidy level. Haploid indicates one copy; diploid, two copies —this is the most common situation in higher organisms; triploid, three copies; tetraploid, four copies; hexaploid, six copies; and polyploid, a general term meaning more than two copies.

2.1.4 Meiosis

During the process of sexual reproduction, the chromosomes need to be separated into distinct sets. This is accomplished by meiosis. This is the process by which the two sets of chromosomes in a diploid organism are separated to produce haploid gametes. An illustration is given in Figures 2.1 and 2.2.

2.1.5 Dominance

Since in most organisms each gene occurs in at least two copies, the alleles at the same locus can have various relationships to each other.

Dominance is the ability of the dominant allele to express itself in the presence of the recessive allele, so that the dominant phenotype is expressed. Biologically this can often be explained by the fact that the recessive allele is essentially non-functional, while the dominant allele produces enough of the particular gene product to satisfy the need of the organism. An example is given in Table 2.1. The other allele is called recessive.

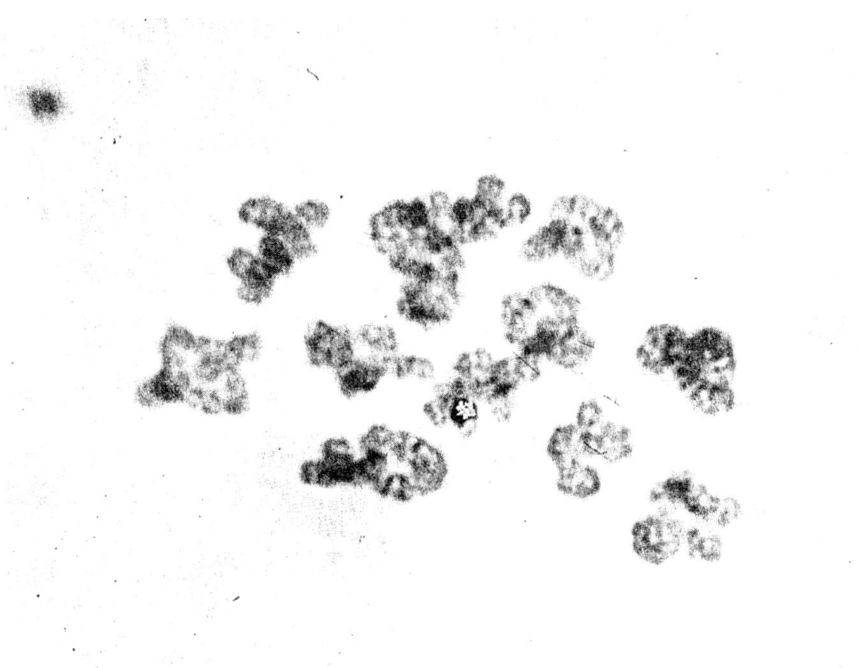

Figure 2.1. First metaphase of meiosis in loblolly pine. All 12 pairs of chromosomes are clearly visible.

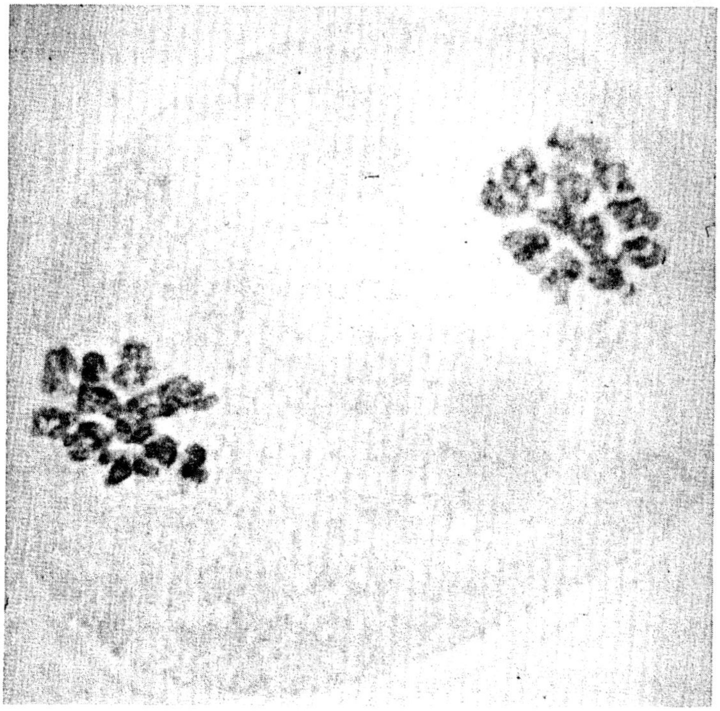

Figure 2.2. Anaphase I during meiosis of loblolly pine. The 12 pairs of chromosomes have separated into two sets.

2.1.6 Epistasis

Interaction among genes at two different loci is called epistasis. This the ability of a gene at one locus to affect the expression of a gene at another locus. This can occur readily if the two genes control two different steps in the same chemical pathway. An illustration is given in Table 2.2. In this example the desirable trait is expressed only if a dominant gene is present at locus A as well as locus B. If either locus is homozygous for the recessive gene the undesirable trait is expressed.

Table 2.1. Expression of dominance for a qualitative and a quantitative trait.

Genotype	Qualitative Trait	Quantitative Trait
AA	Normal	Tall
Aa	Normal	Tall
aa	Albino	Short

2.1.7 Mendel's laws

Mendel discovered that inheritance follows a distinct pattern and formulated two laws describing this pattern. His first law is the law of segregation: traits are inherited as discrete units which cleanly

Table 2.2. Expression of epistasis for a qualitative and a quantitative trait.

Genotype	Qualitative Trait	Quantitative Trait
A-B-	Normal	Tall
aaB-	Albino	Short
A-bb	Albino	Short
aabb	Albino	Short

separate when a new generation is produced. An example is given in Table 2.3. Alleles "A" and "a" clearly separate and recombine at random. There are 3 genotypes possible: AA, Aa and aa. Since genotype Aa can arise in two different ways its frequency is twice the product of the allele frequencies.

Table 2.3. The genotypes and phenotypes produced by segregation and recombination for a qualitative trait at a single locus with two alleles. Male and female parental genotypes are both Aa.

Female Gametes	*Male Gametes* A	a
A	AA	Aa
a	Aa	a

Genotype	Frequency	Phenotype
AA	1	Normal
Aa	2	Normal
aa	1	Albino

Mendel's second law is known as the law of independent assortment: alleles at different loci recombine independently from each other. An example with two loci, two alleles per locus, and equal frequencies of the alleles is shown in Table 2.4. In this example, there are four possible kinds of gametes, each occurring 1/4 of the time. They will combine to give 16 possible zygotes, each occurring 1/16 of the time. Since some of the zygotic genotypes can be formed in more than one way, there are in reality only 9 different genotypes possible, occurring in the frequencies shown in Table 2.4. Let us now assign phenotypes to these genotypes. Qualitative traits are the easiest to visualize. Let us assume that A is a dominant gene for tallness and B is a dominant

SUMMARY

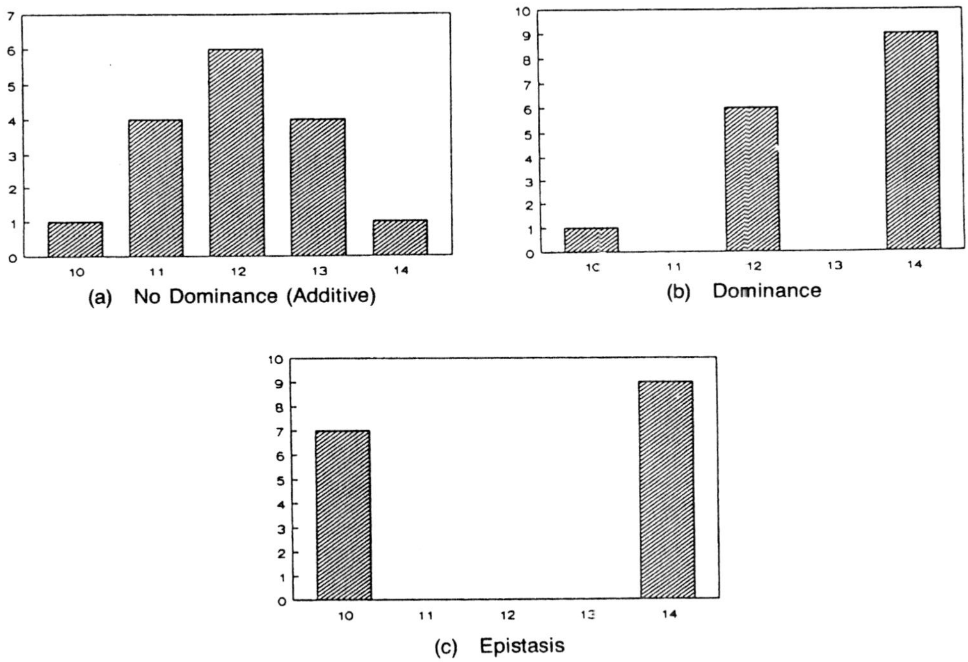

Figure 2.3. Frequency distribution of phenotypes assuming all additive variance (a), the presence of dominance variance (b), and the presence of epistasis (c).

Table 2.4. The genotypes produced by segregation and independent assortment of two heterozygous loci with two allele per locus. Male and female parental genotypes are both AaBb.

Female Gametes	*Male Gametes* AB	Ab	aB	ab
AB	AABB	AABb	AaBB	AaBb
Ab	AABb	AAbb	AaBb	Aabb
aB	AaBB	AaBb	aaBB	aaBb
ab	AaBb	Aabb	aaBb	aabb

Genotype	Frequency
AABB	1
AABb	2
AAbb	1
AaBB	2
AaBb	4
Aabb	2
aaBB	1
aaBb	2
aabb	1
	16

Table 2.5. Frequencies of genotypes for a qualitative trait in a two-locus system under dominance and epistasis.

	Dominance		Epistasis	
Genotype	Phenotype	Frequency	Phenotype	Frequency
A-B-	Tall straight	9	Normal	9
A-bb-	Tall crooked	3	Albino	3 } 7
aaB-	Short straight	3	Albino	3 } 7
aabb	Short crooked	1	Albino	1 } 7

gene for straightness. This will lead to four classes of trees: tall and straight, tall and crooked, short and straight, and short and crooked in the ratio 9:3:3:1 as indicated in Table 2.5. The genotypes are indicated as well in Table 2.5.

Epistasis is somewhat more complicated conceptually, but is very simple in outcome. Let us take a case where there are two dominant genes necessary for the formation of chlorophyll. This is biochemically quite possible, since there are many steps in the synthesis of chlorophyll, each controlled by one gene (resulting in an enzyme). To have normal chlorophyll a tree has to have at least one A and one B allele. All other genotypes are albino. As indicated in Table 2.5, equal frequencies of the dominant and recessive alleles result in a ratio of 9 normal to 7 albino trees.

Quantitative traits are somewhat more involved. Table 2.6 represents a case where only additive effects are present. The trait is 5-yr-height and the base level is 10 ft. Each allele represented by a capital letter contributes 1 additional foot of height. The height, therefore, ranges

Table 2.6. Genotypic and phenotypic frequencies for a quantitative trait in a two-locus model with only additive genetic variation.

Genotypes	Frequency	Phenotype: 5-yr. Height (ft.)
AABB	1	14
AABb	2	13
AAbb	1	12
AaBB	2	13
AaBb	4	12
Aabb	2	11
aaBB	1	12
aaBb	2	11
aabb	1	10

from 10 to 14 feet with the expected ratios shown in Table 2.6. Table 2.7 takes it one step further. When dominance is present, one A or one B allele will increase height by two feet. Adding a second A or B will have no effect. This will give heights of 14, 12 and 10 feet in a 9:6:1 ratio, which follows directly from the dominance column of Table 2.7. When epistasis is present at least one A and one B allele are required to give a height of 14 ft. All other genotypes are 10 ft. A summary of the height distributions of the quantitative cases is given in Figure 2.3.

Table 2.7. Genotypic and phenotypic frequencies for a quantitative trait in a two-locus model with genetic variation involving dominance and epistasis.

Genotype	Frequency	Dominance Phenotype 5-yr Height (ft)	Epistasis Phenotype 5-yr Height (ft)
A-B-	9	14	14
A-bb	3	12	10
aaB-	3	12	10
aabb	1	10	10

2.2 Population Genetics

In this section we discuss a number of concepts, processes and laws that were developed particularly in relation to natural populations, but which to a large extent also apply to populations of domesticated species. We will first focus on the process of selection, which is the process of allowing some parts of the population to contribute relatively more gametes (and thus, genes) to the next generation, compared to their frequencies in the parent population.

2.2.1 Natural selection

Natural selection is nature's way of producing the next generation: the genetic survival of the most prolific. The process of natural selection is most known through the work of Darwin. Those individuals contributing relatively more gametes to the next generation will also contribute relatively more genes to the next generation and thus increase the frequency of these genes in the population.

Under conditions of artificial selection, people will determine which individuals will produce the next generation. This is an essential step in all breeding programs. The probability of success is greatest when artificial selection favors different genes than were favored by natural selection. Artificial selection that parallels natural selection is much less likely to produce much progress. To treat this process mathematically, a number of parameters need to be determined.

2.2.2 Survival rate

The survival rate is the ratio of individuals of the same genotype before and after selection. In population genetics it is usually meant to measure reproductive success. An example is given in Table 2.8.

Table 2.8. Calculation of survival rate.

Genotype	No. of Individuals Before Selection	No. of Individuals After Selection	Survival Rate
AA	2000	1600	.8
Aa	4000	2400	.6
aa	2000	800	.4

The relative fitness of a genotype is the survival rate of a genotype divided by the survival rate of the most fit genotype. The relative fitness is designated by the coefficient $\bar{W}$.

The selection coefficient (s) is the opposite of fitness: s=1-W. It is the proportion of a genotype that is eliminated. The appropriate calculations using the example from Table 2.8 are shown in Table 2.9.

Table 2.9. Calculation of relative fitness (W) and the selection coefficient (s).

Genotype	W	s
AA	.8/.8 = 1	0
Aa	.6/.8 = .75	.25
aa	.4/.8 = .5	.5

2.2.3 Hardy-Weinberg equilibrium

When no changes are taking place in the allele frequencies of a population, it is said to be in Hardy-Weinberg equilibrium. Hardy-Weinberg's law states that, in the absence of selection, mutation, migration and random drift, the allele frequencies remain unchanged and the genotypes AA, Aa, and aa are present in the frequencies p^2, 2pq and q^2 when p and q are the frequencies of A and a.

The fitness values of the three genotypes are W_{AA}, W_{Aa}, and W_{aa} respectively, and $\bar{W}$, the weighted average of the fitness of the three different genotypes, is the average fitness of the population (Table 2.10). This is a very useful concept to study the effect of selection, since selection will always act to increase the average fitness of the population. Figure 2.4 shows a situation in which the homozygous dominant individuals have a fitness of 1, whereas the heterozygotes

Table 2.10. Genotypes, frequencies, fitness values and average fitness of a population with two alleles at a single locus, A and a, with frequencies of p and q respectively.

Genotype	AA	Aa	aa
Frequency	p^2	2pq	q^2
Fitness	W_{AA}	W_{Aa}	W_{aa}
Average Fitness	$\bar{W} = p^2W_{AA} + 2pqW_{Aa} + q_2W_{aa}$		

have an intermediate fitness and recessives have a fitness of .5. The figure shows the average fitness of the population as a function of allele frequencies. In this case selection against the recessive allele takes place and, over a number of generations, the fitness of the population will travel up the slope of the line until it reaches the frequency of p = 1. If other evolutionary forces are present, such as mutation and migration, the frequency of the recessive allele will be changed until it is exactly counter-balanced by mutation and migration.

In Figure 2.5 the heterozygote is depicted as being the most fit. Selection will therefore drive the fitness towards the maximum, which in this case lies at a frequency of p = .6 and q = .4. Once this frequency is reached an equilibrium is established and both alleles will remain in the population. This type of selection, therefore, is called stabilizing selection. The frequencies of the alleles will be skewed towards the homozygote with the greater fitness.

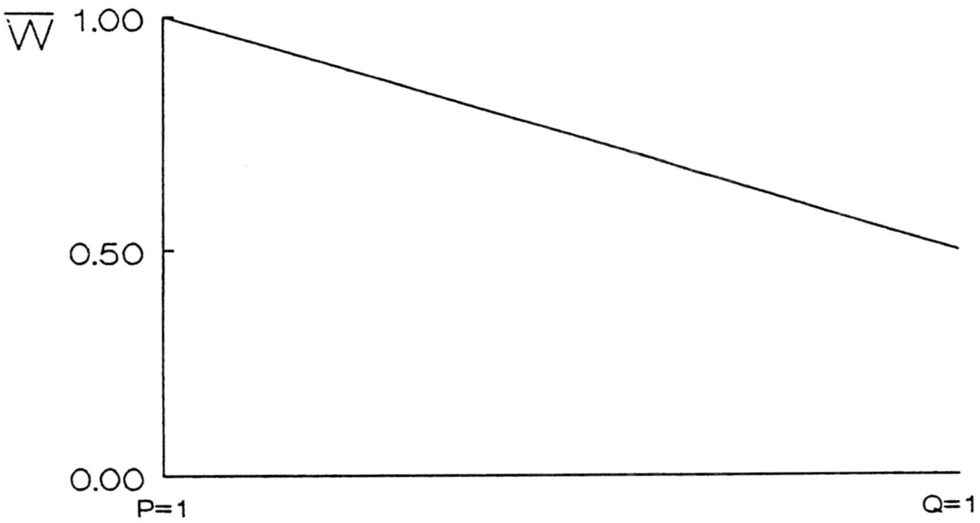

Genotype	W
AA	1
Aa	.75
aa	.50

Figure 2.4. Relation of average population fitness ($\bar{W}$) to allele frequency leading to directional selection.

In Figure 2.6, the heterozygote is less fit than the homozygotes. In this case the situation is unstable. If the starting point is to the left of the minimum, the allele frequency p will go to 1; if it is to the right of the equilibrium the allele frequency q will be driven to 1. This is called disruptive selection.

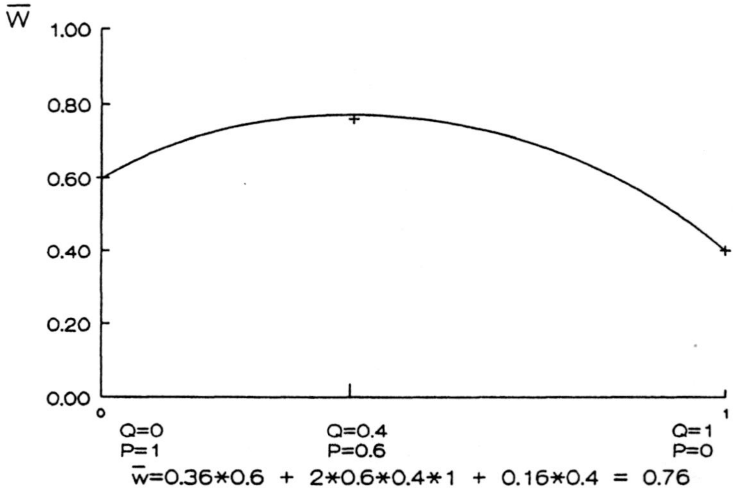

Genotype	W
AA	.6
Aa	1
aa	.4

Figure 2.5. Relation of average population fitness to allele frequency leading to stabilizing selection.

2.2.4 Polymorphism

There are many situations in which two or more genotypes are present in the population in non-trivial frequencies. Such a situation is called a polymorphism. There are various sorts of polymorphisms. A balanced polymorphism is one that is maintained indefinitely in a population by two opposing evolutionary forces.

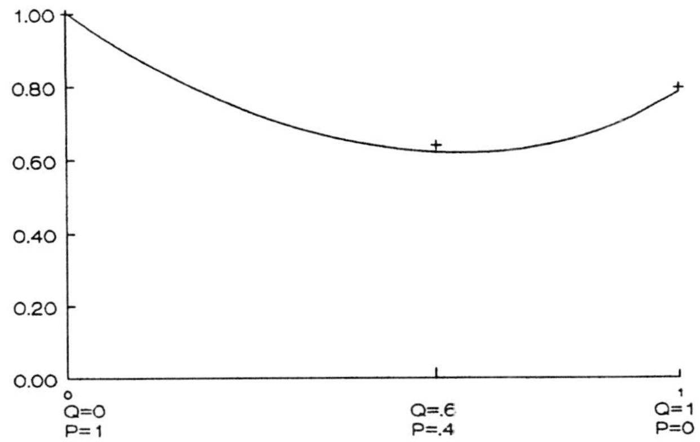

Genotype	W
AA	1
Aa	.4
aa	.8

$$\bar{W}_{min} = .16\times1 + 2\times.24\times.4 + .36\times.8 = .64$$

Figure 2.6. Relation of average population fitness to allele frequency leading to disruptive selection.

2.2.5 Balance mechanisms

There are several possible mechanisms that can maintain polymorphisms in a population. Following are a few prominent ones:

Mutation and selection: Balance exists when selection against a recessive allele removes the allele as fast as it is replaced by mutation.

Gene flow (migration) and selection: Balance exists when gene flow replaces an allele as fast as selection removes it.

Frequency dependent selection: This occurs when an allele is selected for when it is rare, but selected against when it is common. As a result it may stabilize at an intermediate frequency. This situation results quite often from predation.

Stabilizing selection: Selection for the heterozygotes is balanced by selection against the homozygotes resulting in both alleles remaining in the population at an intermediate frequency.

2.2.6 Effect of natural selection

The end result of natural selection is the development of geographical races (provenances) and altitudinal races that are closely adapted to their local environment. Of course eventually this could lead to the formation of new species. Our interest, however, is primarily in the development of races within the species. This makes it necessary to develop carefully designed rules for seed transfer. Seed transfer is not necessarily bi-directional; in southern pines, for instance, seed can be readily moved from west to east and south to north, but the opposite movements can often lead to rather disastrous results. An illustration of geographic races is given in Figure 2.7.

2.2.7 Concepts of importance for seed transfer

When designing rules for seed transfer it is necessary to keep a number of concepts clearly in mind. Some of these are just being introduced. One of the concepts is that of a seed zone. This can be defined as a relatively homogeneous physiographic region in which seed can be freely moved. An example is given in Figure 2.8, which depicts the seed zones adopted by the Western Gulf Forest Tree Improvement Program (WGFTIP).

Since seed movement is not necessarily bi-directional we need two other concepts: seed deployment zone and seed procurement zone.

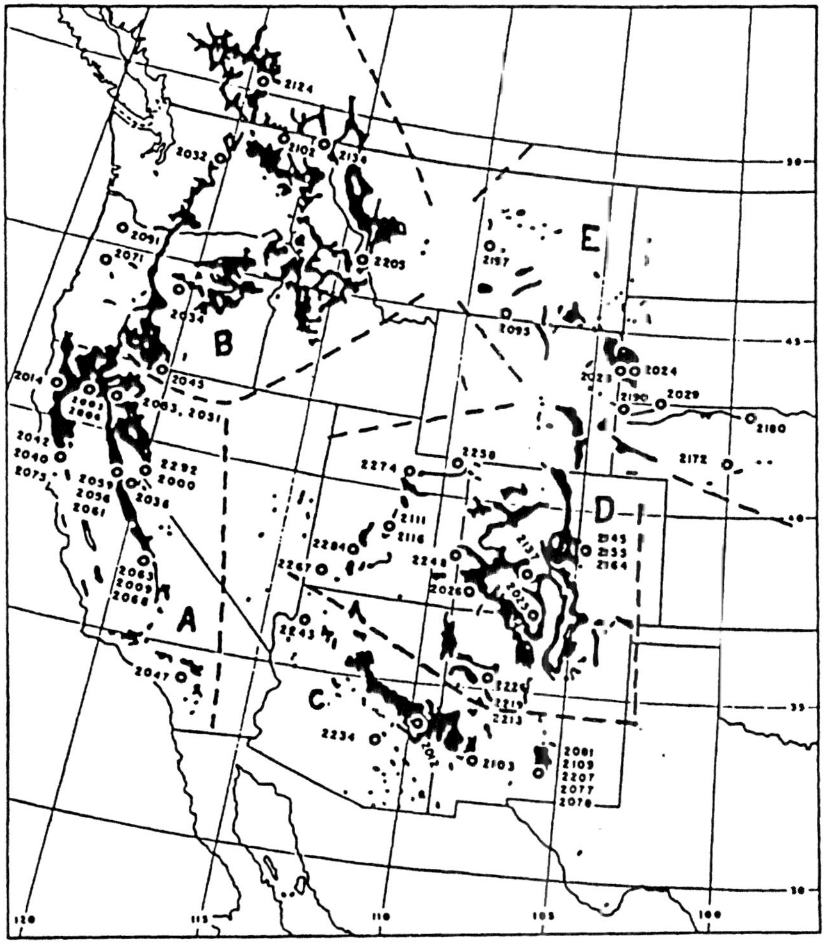

Figure 2.7. Ecotypic division of ponderosa pine progenies grown for two years at East Lansing. A = California, B = Northern Plateau and Willamette Valley, C = Southern Interior, D = Central Interior, E = Northern Interior. Numbers show the location of the stands that were sampled. (Distribution map by E. L. Little Jr., U.S. Forest Service.) (Wells 1964).

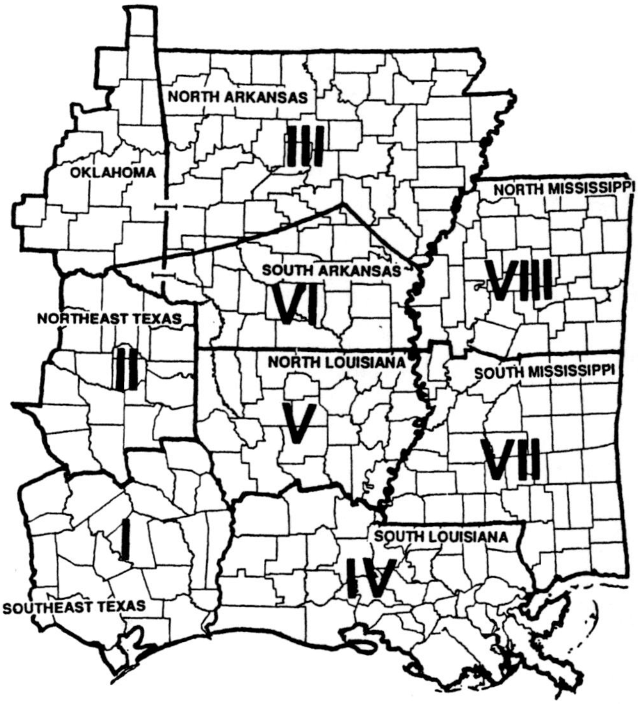

Figure 2.8. Seed zones adopted by Western Gulf Forest Tree Improvement. The seed deployment zone is the zone in which seed from a particular seed zone can be used.

The seed deployment zone is the zone in which seed from a particular seed zone can be used. Figure 2.9 depicts the recommendations of Wells (1969) for the movement of loblolly pine seed. The seed from Southeast Texas can be collected there, but can be planted all through Area 1, while seed collected in Livingston Parish can be planted throughout Area 2. In this case, Areas 1 and 2 are the deployment zones for these particular provenances. The counterpart of the deployment zone is the procurement zone. It is the area from which

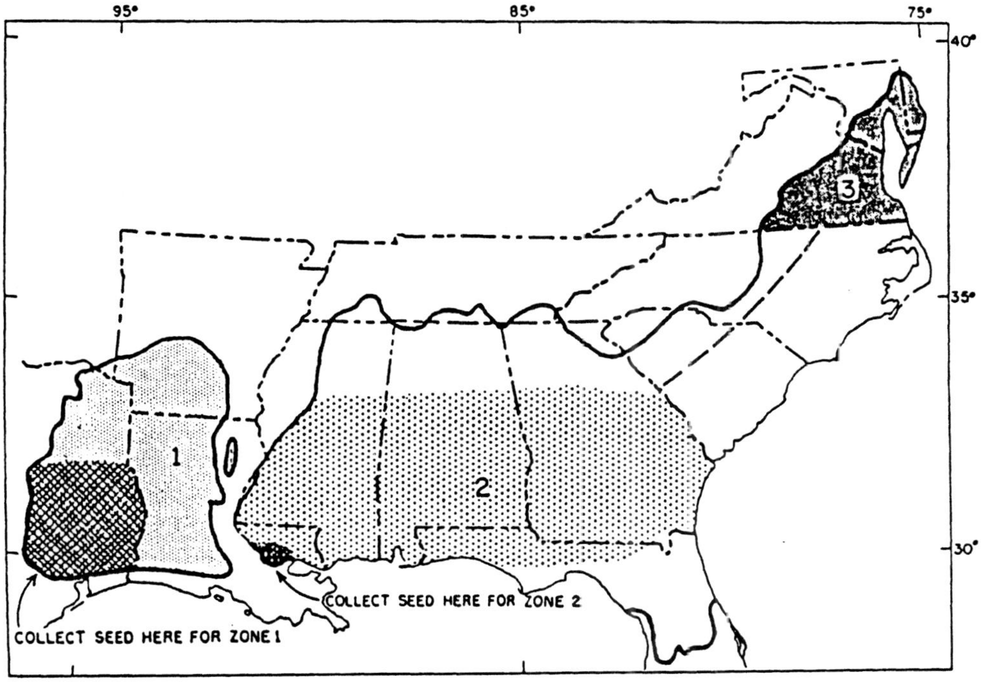

Figure 2.9. Suggested seed collection and planting zones for loblolly pine. Natural range outlined by solid black line (Wells 1969).

seed can be procured for planting in a particular seed zone. An illustration is given in Figure 2.10. In South Arkansas, seed can be used from South Arkansas itself, as well as from Northeast Texas and North Louisiana. One cannot necessarily, however, move the North Arkansas seed south.

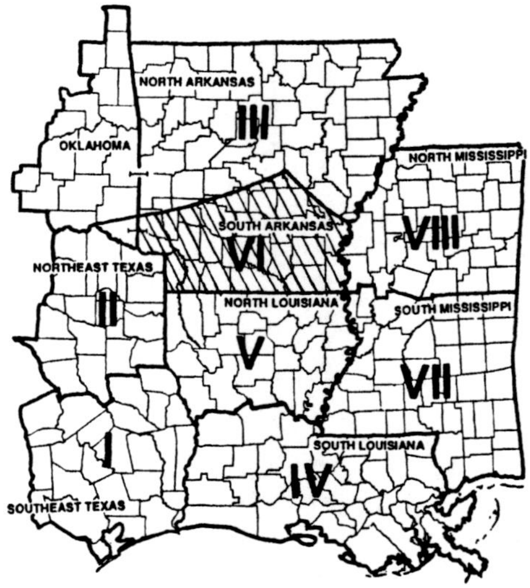

Figure 2.10. The procurement zone for WGFTIP seed zone VI is comprised of seed zones II, V and VI.

2.3 Quantitative Genetics

In this section we will define the terms used in quantitative genetics, outline the procedures for calculating parameters and describe their use in selection and estimating genetic gain.

2.3.1 Single trait. One location.

One of the fundamental parameters used to measure variation is the standard deviation (σ). It is computed using the following steps.

1) Square all deviations from the mean (Figure 2.11)
2) Add them together
3) Average them
4) Take the square root of the average.

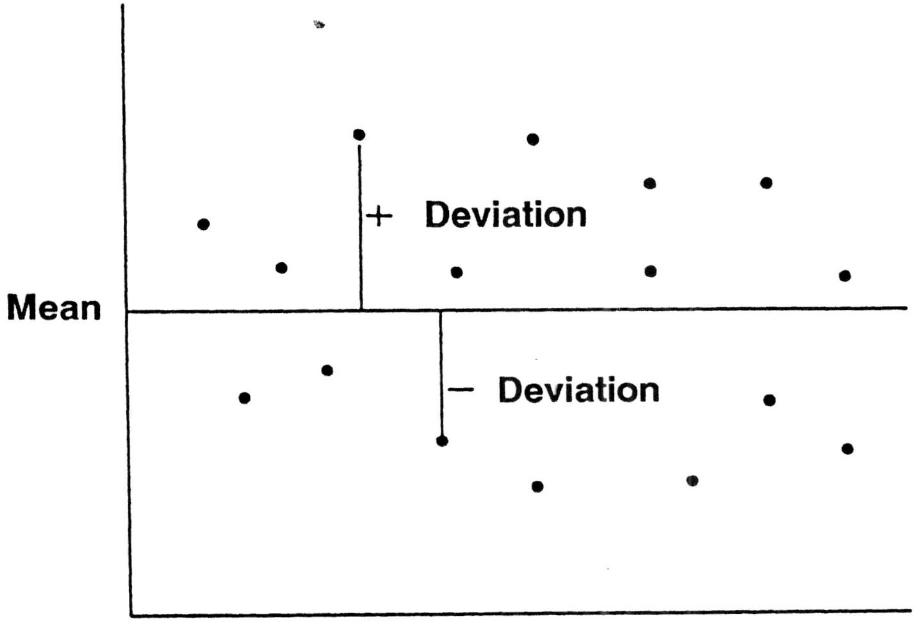

Figure 2.11. Schematic representation of mean and standard deviation.

Another measure is the variance (V). It is the square of the standard deviation. The variance has the interesting mathematical property that it can be broken down into portions (components) associated with different causes of variation. The process of doing this is appropriately enough called an *analysis of variance.*

These variance components - the portions of variance associated with different causes of variation - depend very much on the nature of an experiment. For geneticists, the following components are of particular interest.

Phenotypic variance (V_p): This is the total variance among all phenotypes in the population.

Genetic variance (V_g): This is the variance associated with the genetic differences among all trees in the population.

Additive genetic variance (V_a): This is that portion of the genetic variance associated with additive genetic effects. Genetic effects are called additive to the extent that progenies are exactly intermediate between the parents.

Non-additive genetic variance (V_{na}): This is the portion of the genetic variance associated with non-additive genetic effects. Genetic effects are called non-additive to the extent that progenies are not exactly intermediate between the parents. This can be caused by dominance and epistasis.

Environmental variance (V_e): This is the portion of the variation associated with environmental effects only.

We now have the following relationships:

$$V_p = V_g + V_e$$
$$V_g = V_a + V_{na}$$
$$V_p = V_a + V_{na} + V_e$$

2.3.1.1 Heritability

The components described above give us the opportunity to calculate the parameter measuring the strength of the inheritance. This is called the heritability, which is the ratio of genetic variance to phenotypic variance. The variance components described above allow us to calculate a number of different heritabilities. *Narrow sense heritability* (h^2) is the ratio of the additive genetic variance to the phenotypic variance (V_a/V_p). The *broad sense (gross)* heritability (H^2) is the ratio of total genetic variance to the phenotypic variance (V_g/V_p). *Individual tree heritability* (h_i^2) is the heritability based on individual tree measurements, and is the most common usage of heritability. *Family heritability* (h_f^2) is the heritability based on family averages. When this is based on half-sib family averages, the genetic variance is 1/4 V_a. The phenotypic variance to be used here is the variance among the family means. This can be calculated quickly from an analysis of variance table by dividing the family mean square by the number of trees per family.

These heritabilities have to be interpreted carefully to be correct. Each heritability estimate is specific to the population, the trait, and the environment on which the estimate is based. The heritability is not specific, for instance, to the species.

The narrow sense heritability is used most appropriately for seed orchards and seed production areas, that is, for species and programs in which sexual recombination occurs, and additive effects are most important.

The broad sense heritability, on the other hand, which includes all genetic variation, is particularly appropriate for vegetatively propagated material. Because of this, the gains obtained by vegetative propagation are generally much larger than those obtained from sexually propagated material.

One way to increase heritability is to reduce the environmental variance. If a progeny test is planted on a variable site, the site is poorly prepared and weeds and hardwood sprouts are not controlled, the environmental variance can be quite high. As a result observed differences will be largely environmentally caused and the heritability will be very low. This is the reason for the emphasis placed on careful site selection, and proper layout and maintenance of progeny tests as described in Chapter 4.

Heritabilities generally have large errors associated with them. In view of this, the consistency of the values reported in the literature is actually rather surprising.

2.3.1.2 Numerical example

A numerical example using the quantitative model used in Tables 2.7 and 2.8 is given in Table 2.11. Appropriately enough, the additive model showed no dominance variation. In both cases, no environmental variation was found because it was not originally included in the model.

2.3.1.3 Selection and genetic gain

When many trees are selected for a particular trait, the average of this trait for the selected trees will be different from that of the original population. The difference between the average of the selections and the population average is called the *selection differential* (S). For instance, if the average population height is 80 feet and the average height of the selected trees is 100 feet, then the selection differential

Table 2.11. Numerical example of quantitative genetic variation using the genetic model from Tables 2.7 and 2.8.

Genotypes	Number of Individuals	Additive 5-Year Height	Dominance 5-Year Height
AABB	1	14	14
AABb	2	13	14
AAbb	1	12	12
AaBB	2	13	14
AaBb	4	12	12
Aabb	2	11	12
aaBB	1	12	12
aaBb	2	11	12
aabb	1	10	10

Additive Variation Only	Additive Plus Dominance Variation
$V_a = 1.0$	1
$V_{na} = 0$	.5
$V_e = 0$	0
$h^2 = 1.0$	$h^2 = .7$
$H^2 = 1.0$	$H^2 = 1.0$

is 20 feet. Since variability differs greatly from trait to trait, the absolute difference in value does not give a good indication of the amount of selection pressure applied. One therefore, commonly uses another measure, the *selection intensity* (i), which is the selection differential divided by the phenotypic standard deviation (σ_p). The standard deviation provides a common scale unit, which is particularly handy when comparing different traits, or when dealing with several traits at the same time[2]. An example is given in Table 2.12, which shows that when dealing with two traits, two rather different selection differentials can be very similar in selection intensity.

The parameters discussed so far now enable us to calculate the *genetic gain* (ΔG). The *predicted gain* can be computed as follows:

$$\Delta G = h^2 S = i\, h^2\, \sigma_p$$

The gain that can be obtained by selection in a given population is therefore a function of the amount of variation present (σ_p), how much of this variation is genetic (h^2), and how intensively one selects in this population (i).

The actual gain obtained is often less than the predicted gain. Table 2.13 gives some examples of the expected gain computation. It is often convenient to express the expected gain in terms of a percent. Some parameters are related to the quality of the selections, and once progeny test information becomes available, they can be calculated to

---

[2]*Editor's note: Lerner (1958) in "The Genetic Basis of Selection" defined selection intensity simply as "the percentage of the population permitted to reproduce itself". The definition given here is equivalent to Lerner's standardized selection differential ($\bar{i}$), which Falconer (1981) called "the intensity of selection". This discrepancy in terminology can lead to confusion, and suggests that, as van Buijtenan has, the writer clearly indicate which definition is being used.*

Table 2.12. Calculation of selection differential and selection intensity for height and diameter in *Hypoplatanus obtusa* vB.

Trait	Population Average	Avg. of Selected Trees	S	Standard Deviation	i
Height (ft)	80	100	20	10	2
Diameter (in)	20	30	10	5	2

evaluate potential parents. The *general combining ability effect* (gca) is the additive genetic contribution of a selected parent to its progeny. The gca of an individual is therefore one-half of its superiority over the population average. A good general combiner, on the average, has good progeny regardless of the other parent. A good general combiner is expected to possess many genes with positive additive contributions to its breeding value.

Specific combining ability (sca) is the deviation of a specific combination of two parents from the sum of their gca effects. A cross showing good specific combining ability is better than expected. In practice breeders actually consider the combined result of gca and sca in producing an outstanding specific cross. In trees the major contribution usually results from starting with two outstanding parents.

Using the dominance model: AAbb (12) and aaBB (12) would show specific combining ability because all progeny would be AaBb (14).

Table 2.13. Calculation of expected gain in a population of *Hypoplatanus obtusa* vB.

Trait	h^2	S	σ_p	i	ΔG	Population Average	% Gain
Height	.2	30	10	3	6	80	7.5
Diameter	.1	15	5	3	1.5	20	8.0

The *breeding value* is closely related to general combining ability and is the net result of all the additively acting genes affecting a particular trait. A tree having a high breeding value for a particular trait would likely have many genes with positive contributions. It should also be a good general combiner.

"If an individual is mated to a number of individuals taken at random from the population, then the breeding value is twice the mean deviation of the progeny from the population mean" (Falconer 1981). An example is given in Table 2.14, which shows that using the additive model, the breeding value of genotype of AABB would be two while the double recessive would be minus two. This definition of Falconer's is not quite correct. The results should be multiplied by the family heritability to give the correct answer.

2.3.2 Single trait. Multiple locations.

When multiple locations or sites are used in genetic testing programs, a new concept comes into play: the *genotype by environment interactions* (GxE). If the same families, when grown in different environments, do not rank in the same order, important and potentially

Table 2.14. Calculation of breeding values using the example of additive genetic variation from Table 2.11.

Genotype	Breeding Value
AABB	2
AaBb	0
aabb	-2

troublesome GxE interaction exists. Commonly one finds that the best and the worst families do not change places, but that there is a fair amount of switching in the middle range. The illustrations in Figure 2.12 graphically show the various types of interaction. The individual family heights are plotted against the average plantation height to produce the corresponding curves. In Figure 2.12a, where the lines are completely parallel, absolutely no interaction is present. In Figure 2.12b, where the lines diverge, the analysis of variance will show interaction. However, the interaction is trivial since no rank changes are taking place and the same families will be selected at all locations. When the lines cross, however, there is interaction present which needs to be considered very seriously. In this case one family performs the worst at the poor sites, but performs best at the very good sites. In such a case, the breeder can either try to find genotypes that perform well on a broad range of sites, which are generally referred to as stable genotypes, or the breeder can try to match specific genotypes to specific sites. The latter option may produce more genetic gain, but it is also far more costly.

2.3.3 Multiple traits

When describing variation among multiple traits, the term *covariance* (Cov_{12}) replaces the variance concept. Instead of dealing with squared deviations from the means, one deals now with cross-products of deviations from the mean as shown in Figure 2.13. The covariance is an average cross-product. If both deviations are generally in the same direction the covariance is positive as in Figure 2.13c; if both deviations generally are in opposite directions the covariance is negative (Figure 2.13b). If there is no relation, the covariance is close to zero (Figure 2.13a). If the relation is perfect, all points fall on a straight line (Figure 2.13d). The covariance concept is a generalization of the variance concept, and can be partitioned into components just like the variance. The covariance of a trait with itself is the variance of that trait.

The *correlation* (r_{12}) is a standardized version of the covariance obtained by dividing the covariance by the product of the standard deviations of the two traits involved, as follows:

$$r_{12}=\frac{Cov_{12}}{\sigma_1\sigma_2}$$

As a result of this manipulation the range of the correlation coefficient is expected to be from -1 to +1. The procedure followed here is more or less analogous to the way selection differential was expressed in terms of standard deviations to obtain the selection intensity. By doing this, relationships among various traits can be put on a common scale.

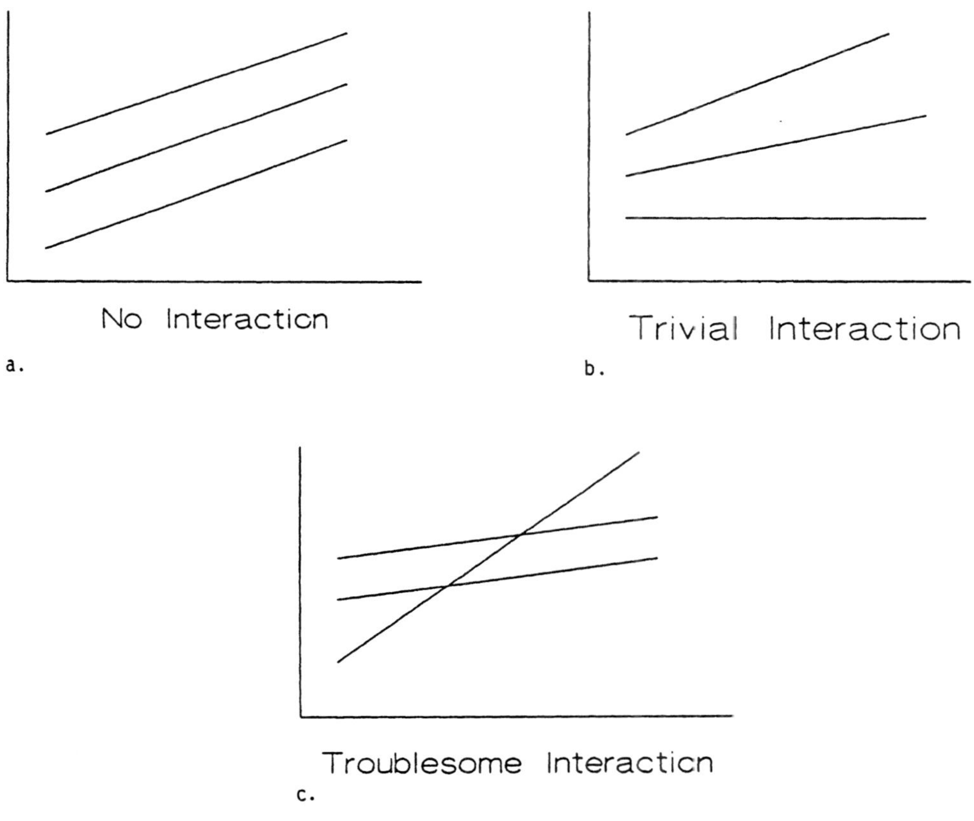

Figure 2.12. Different types of genotype by environment interaction. In all cases, average family height at a location is plotted against the overall average height at that location.

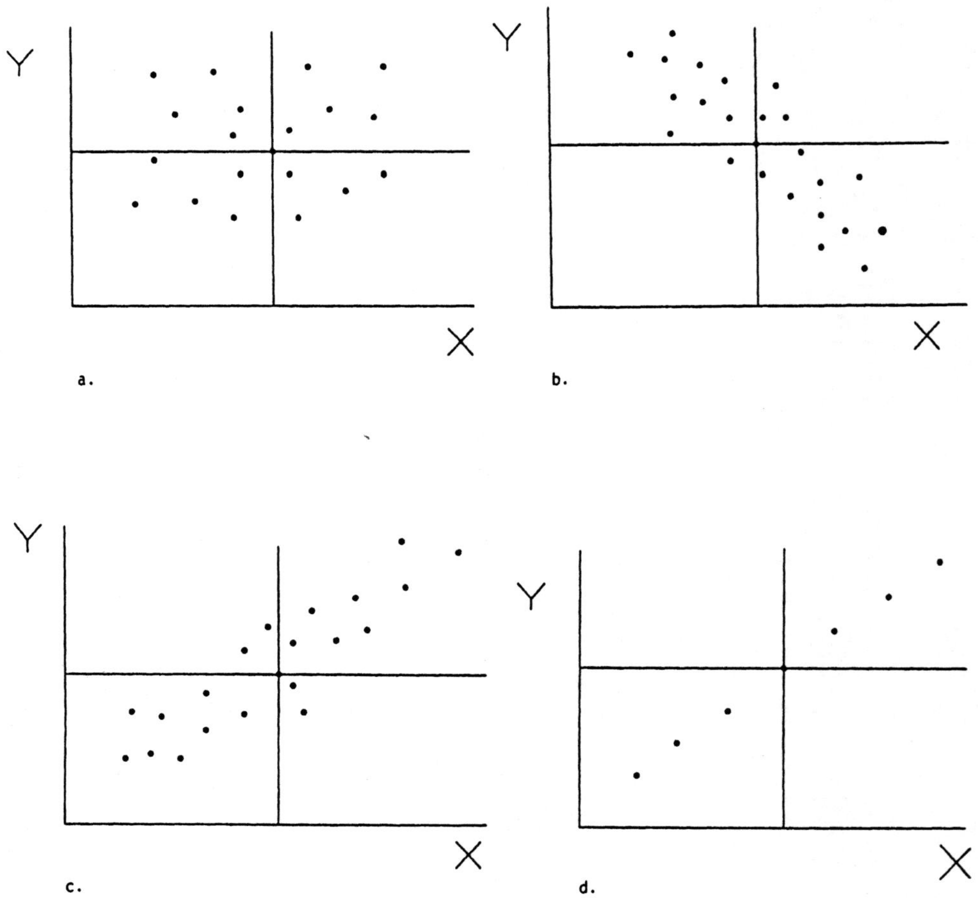

Figure 2.13. Distribution of points for two variables leading to different correlations. a) No correlation b) positive correlation c) negative correlation d) perfect correlation.

The *phenotypic covariance* ($Cov_{p_1p_2}$) is the "average" cross product between the phenotypic values of trait 1 and trait 2 for each individual. It includes all genetic variation as well as the environmental variation. It can also be expressed as a correlation, the *phenotypic correlation* ($r_{p_1p_2}$). If it is strongly positive (close to 1) the two traits usually vary in the same direction. A good example is height and diameter.

The *genetic covariance* ($Cov_{a_1a_2}$) is analogous to the genetic variance, but consists of the component of the cross-products related to the genetic differences among families. The *genetic correlation* ($r_{a_1a_2}$) is the correlation obtained by dividing the genetic covariance by the product of the genetic standard deviations. These in turn can be obtained by taking the square root of the product of the genetic variances of the two traits. Therefore:

$$r_{a_1a_2} = \frac{Cov_{a_1a_2}}{\sqrt{V_{a_1}V_{a_2}}}$$

The genetic correlation is expected to range from -1 to +1, but occasionally exceeds these values, because of errors in the estimation of the variances and covariances. A positive correlation means the two traits are usually associated with each other, a negative correlation means that one trait tends to go down as the other goes up as a result of genetic manipulation.

In a number of cases volume growth and wood specific gravity behave this way: a genetic increase in growth is often associated with a decrease in specific gravity.

The genetic correlation per se is not a very useful statistic. It can be useful in combination with some other parameters, allowing one to predict the response of trait 2 because of selection for trait 1 as follows:

The correlated response in trait 2 = $i\ h_1 h_2\ r_{a_1 a_2}\ \sigma_{p_2}$ where h_1 and h_2 are the square roots of the heritabilities of trait 1 and trait 2.

The correlated response can be calculated more elegantly using the *coefficient of genetic prediction* (CGP). CGP_{12} is a coefficient predicting the response in trait 2 by selecting for trait 1, and is calculated as:

$$CGP_{12} = \frac{Cov_{a_1 a_2}}{\sigma_{p_1} \sigma_{p_2}}$$

CGP is a generalization of the heritability concept, and the CGP of a trait with itself is the heritability of that trait. The CGP is an elegant means for computing a variety of responses:

a) The correlated response to selection as described above.

b) One can consider a trait measured at two ages as two different traits, such as 5-year height and 25-year height. The CGP for these two traits is the proper "juvenile-mature correlation" to use for predicting the response at age 25 by selecting for height at age 5.

c) One can consider a trait measured at location 1 and at location 2 as two different traits. This can be used as a means of coping with GxE interactions. It allows one to predict the gain at location 2 obtained by selecting on the basis of measurements made at location 1.

The *correlated response* is the response in trait 2 as a result of selection in trait 1. Using the coefficient of genetic prediction, it is calculated as follows:

$$\text{Correlated response in trait 2} = i \, CGP_{12} \, \sigma p_2$$

If trait 1 and trait 2 are the same, this is identical to the predicted gain formula discussed earlier. Some examples are given in Tables 2.15 and 2.16. Table 2.15 shows increased survival is somewhat as-

Table 2.15. Heritabilities (on the diagonal) and coefficients of genetic prediction from the combined locations analysis for seven plantings of two-year-old longleaf pine in the Western Gulf region.

Trait	Survival	Grass Stage Emergence	Brown-Spot Severity Score
Survival	0.56	0.18	0.07
Grass Stage Emergence			0.05
Brown-Spot Severity Score		0.54	0.74

Table 2.16. Heritabilities (on the diagonal) and coefficients of genetic prediction for grass stage emergence in two-year-old longleaf pine across different test locations in the Western Gulf region.

Test	Stone, MS	Pearl River, MS	Tyler, TX	Vernon, LA
Stone, MS	0.37	0.51	0.48	0.34
Pearl River, MS		0.23	0.25	0.14
Tyler, TX			0.40	0.37
Vernon, LA				0.22

associated with earlier emergence from the grass stage, but is not too closely related to brown-spot severity. Similarly, brown-spot severity is not closely related to emergence from the grass stage. This is somewhat surprising and conflicting with other reports from the literature. It is also clear that the heritability of all three traits is quite high. The coefficient of genetic prediction relating the performance of grass-stage emergence at four different locations show that the performance at Stone, Mississippi, Tyler, Texas and Vernon, Louisiana are fairly consistent, while the performance at Pearl River, Mississippi follows a somewhat different pattern. Table 2.17 shows various covariances and correlations obtained from a five-year-old open-pollinated loblolly pine progeny test of the Texas Forest Service.

Table 2.17. Covariances, correlations and coefficients of genetic prediction calculated from a five-year-old open-pollinated loblolly pine progeny test of the Texas Forest Service.

	5-Yr. Height	5-Yr. Diameter	Specific Gravity
	Phenotypic Covariance		
5-Yr. Height	.3907	.2540	-.001252
5-Yr. Diameter		.2446	-.001558
Specific Gravity			.0006514
	Genetic Covariance		
5-Yr. Height	.04829	.03491	-.001589
5-Yr. Diameter		.03342	-.001308
Specific Gravity			.002038
	Phenotypic Correlation		
5-Yr. Height	1	.82	-.08
5-Yr. Diameter		1	-.12
Specific Gravity			1
	Genetic Correlation		
5-Yr. Height	1	.87	-.51
5-Yr. Diameter		1	-.50
Specific Gravity			1
	Coefficient of Genetic Prediction		
5-Yr. Height	.80	.71	-.44
5-Yr. Diameter		.84	-.45
Specific Gravity			.94

Literature Cited

Falconer, D. S. 1981. *Introduction to Quantitative Genetics*. Second Edition. Longman Press, New York. 340 p.

Lerner, I. M. 1958. *The Genetic Basis of Selection*. John Wiley & Sons, Inc., New York. 298 p.

Strickberger, M. W. 1976. *Genetics*. Second Edition. The MacMillan Co. New York. 914 p.

Wells, O. O. 1964. Geographic Variation in Ponderosa Pine. *Silvae Genetica*. 13(4):89-103.

Wells, O. O. 1969. "Results of the Southwide Pine Seed Source Study Through 1968-69." *Proceedings of the 10th Southern Forest Tree Improvement Conference*. pp. 117-129.

Chapter 3
Mating Designs

F. Bridgwater[1]

A breeding strategy is a plan to achieve the objectives of breeding programs. The choice of breeding strategy is largely determined by the kind of gene action predominant in the population of interest, and on the methods that can be used to economically produce large numbers of propagules for plantation establishment. If gene actions are largely additive and seeds can be produced cheaply in wind-pollinated seed orchards, a recurrent selection program to increase the frequency of desirable alleles in the population may be the breeding strategy. If non-additive gene actions are an important part of the total genetic variation, multiple breeding populations to increase the proportion of good specific combinations may be considered.

The elements of a breeding strategy under the breeder's control are: (1) the mode of reproduction, (2) the use of genetic testing, (3) the management of co-ancestry, and (4) the choice of mating design. Thus, the choice of mating design is only one, but an integral factor influencing the choice of breeding strategy; and its impact on the other determinants of breeding strategy must be considered. The

---

[1]*Research Geneticist, Southeastern Forest Experiment Station, USDA Forest Service, and Professor of Forestry, North Carolina State University*

L. Fins et al. (eds.), Handbook of Quantitative Forest Genetics, 69–95.

interactions among these factors have been discussed elsewhere (van Buijtenen and Bridgwater 1986, Bridgwater and Franklin 1985).

3.1 Bases for Comparing Mating Designs

Mating designs are usually chosen to accomplish more than one of the following objectives.

1) Provide information to evaluate parents, i.e., progeny testing. Mating designs may be chosen to provide reliable estimates of general or specific combining ability effects (gca and sca, respectively), or both. Reliable estimates of gca depend on ensuring that each parent is crossed with a sufficient sample of other parents to estimate its potential. Four to five matings per parent are usually considered sufficient for reliable estimates of gca in southern pines (Namkoong 1965). Estimating sca effects depends upon keeping complete records of pedigrees, which implies some type of controlled-cross mating design.

2) Provide for estimates of genetic gains. A mating design must also provide reliable estimates of gca and/or sca if realized genetic gains are to be estimated by measuring the deviation from a standard such as a commercially available seed lot. While the number of parents required to accomplish this is small (see 1 above), many progeny per parent are required to accurately estimate parental potential (Chapter 4). If reliable estimates of expected genetic gains from future progeny of selected parents are desired, mating designs must include large numbers of parents mated in patterns that permit reliable estimates of additive and non-additive genetic variances and co-variances. If both of these objectives must be accommodated in a single mating design and resources are limited, compromises must be made between producing large numbers of progenies per cross and mating many parents. As before, the need for estimating sca effects or variances implies the need for controlled-cross matings and keeping complete records of pedigrees.

3) *Provide estimates of genetic parameters.* Mating designs selected primarily for this purpose must include large numbers of parents for reliable estimates of population variances and co-variances.

4) *Produce a base population suitable for successive cycles of selection and mating.* To satisfy this objective a mating design must first permit co-ancestry management, i.e. use controlled-matings and keep complete pedigree records. Secondly, the selected design must produce sufficient numbers of families and individuals to maintain desired selection differentials. Accomplishing this second goal is problematical when parents of unknown breeding value are mated to provide a base population for the next cycle of selection. In this case, it is necessary to make sufficient numbers of crosses per parent so that there is a reasonable probability that it will have been mated with at least one average or better parent. Thus, it may be possible to select progenies only from parents with good breeding values rather than from matings between parents of good and poor breeding values.

5) Must be cost-effective. The cost of mating designs is determined by the choice of pollination method, the need to keep complete pedigree records, and the complexity of the mating design.

A brief description of several mating designs in use by forest tree breeders follows. Each is evaluated on the basis of the 5 criteria enumerated above.

3.2 Evaluation of Mating Designs

There are two mating designs in general usage that do not provide information on male pedigrees, rather only the female pedigree is known. These are wind-pollination and polycrosses.

3.2.1 Open- or wind-pollinated matings.

Wind-pollinated seeds may be collected from female parents in natural stands or from seed orchards if inference is to be made only to that selected subset of a population.

Progeny testing: Provides for good estimates of gca effects, but no estimate of sca effects since the male parents are unknown.

Gain estimation: Provides for good estimates of realized gains and expected gains that arise from additive genetic variation.

Parameter estimation: The component of variance or covariance for females gives reliable estimates of 1/4 of the additive genetic variance or covariance. Some workers use 1/3 as a multiplier for these estimates since the coefficient 1/4 assumes that male and female parents are not related to each other and that each progeny arises from a different male parent. These assumptions may not be correct. The biology of the southern pines, e.g. suggests that the 1/3 multiplier may be more accurate. Namkoong (1966) examined the effect of varying (1) the relatedness of seed parents with pollinators, (2) the relatedness of pollinators to one another, and (3) the number of effective pollen parents per female. For family sizes of 100, which is common in genetic tests of forest trees, the expected proportion of full-sib pairs was 0.09 for 10 effective pollen parents. At those levels, the proportions of additive and non-additive variances ranged from 0.39 and 0.14, respectively, for male and female parents that were half-sibs, to 0.27 and 0.02, respectively, for unrelated male and female parents. Assuming some unknown mixture of matings ranging from unrelated to mating among half-sibs, the proportion of additive variance estimated by the variance among progenies of seed parents would be between 0.27 and 0.39; i.e., about 1/3.

Base population: Does not provide for control of male pedigrees. If there is great variation for general combining ability among males and inbreeding depression is present, selection in base populations produced from wind pollinations will reduce expected gains (Zobel et al. 1972).

Cost: This is the least costly of mating designs since it requires only seed collection.

3.2.2 Polycross matings

Polycross matings are artificial pollinations with mixtures of male pollens unrelated to any of the candidate females or to each other. It may be desirable to use polycross matings rather than wind pollinations if it is suspected that most wind-pollinated seeds had fewer male parents than desired to estimate general combining abilities of female parents. It may also be desirable to use a mix of pollens selected on some basis to progeny test females. For example, Byram and Lowe (1987) found that testing slash pine female parents for resistance to fusiform rust with a polymix of rust-susceptible males ranked females significantly better than polymix testing with rust-resistant males. They asserted that the rust-susceptible polymix was best because rust-resistance was conferred largely by dominant alleles, and that the absence of these dominant alleles in the females was masked by resistance conferred by resistant males.

Progeny testing: Provides for good estimates of gca effects with sufficient numbers of effective males, as before. Since it has been demonstrated that there is no significant reproductive bias with 9 pollens in equal mixtures on loblolly pines (Wiselogel and van Buijtenen 1988), it may be prudent to use no fewer than that number for that species. As with wind pollinations, no estimate of sca effects are possible since the male parents are unknown.

Gain estimation: Provides for good estimates of realized gains and expected gains that arise from additive genetic variation.

Parameter estimation: The component of variance or covariance for females gives reliable estimates of a proportion of the additive genetic or covariance. The proportion (multiplier) should be adjusted according to the numbers of effective males in the pollen mix, and the co-ancestry of males and females using the tables provided in Namkoong (1966).

Base population: Does not provide for control of male pedigrees if a single polymix is used on all parents. If there is strong variation for general combining ability among males and inbreeding depression is present, selection in base populations produced from polycross matings will reduce expected gains since most selections may be progenies of the same few pollen parents. (Zobel et al. 1972). Subdividing a population into groups of nested polycrosses offers some control over male pedigree at a lower cost than other mating designs that control pedigrees of both sexes.

Three types of ***nested polycrosses***, complete, incomplete, and overlapping have been described (Burdon & Shelbourne 1971). Complete nesting uses all parents as male and female parents, relying upon inherent mechanisms to mitigate against the success of self-fertilizations. The degree of control over the male pedigree depends directly on the number of nests since pollen parents are not known within a nest. Incomplete nesting requires fewer crosses because a sample of males is crossed to the remainder of individuals in the nest. Thus, in addition to the limitations of complete nesting, incomplete nesting provides no estimates of general combining ability for male parents. Overlapping nests use pollen from nest 1 to pollinate nest 2;

pollen from nest 2 to pollinate nest 3; and so forth, until pollen from the last nest is used to pollinate nest 1. Overlapping nests provide estimates of general combining ability for all parents, and eliminates the possibility of self-fertilization. However, since progenies in each nest are related to progenies in another, the ability to ensure out-crossing among progenies from different nests is halved each generation.

Cost: This is only slightly more costly than collecting wind-pollinated seeds if isolation of female strobili from wind-borne pollen is not necessary. If it is desired to limit male parentage to those in the polymix, the additional cost of isolating female strobili must be incurred; but the numbers of crosses required to accomplish the objectives of polycross matings is usually several times smaller than the numbers required for controlled-cross mating designs.

The remainder of the mating designs to be discussed are complete pedigree designs, i.e., they permit maintenance of both male and female pedigrees.

3.2.3 Single-pair matings

Single-pair matings are made among pairs of parents that are mated only once, i.e., without replacement.

Progeny testing: Provides good estimates of full-sib family performance only. Gca and sca effects are confounded.

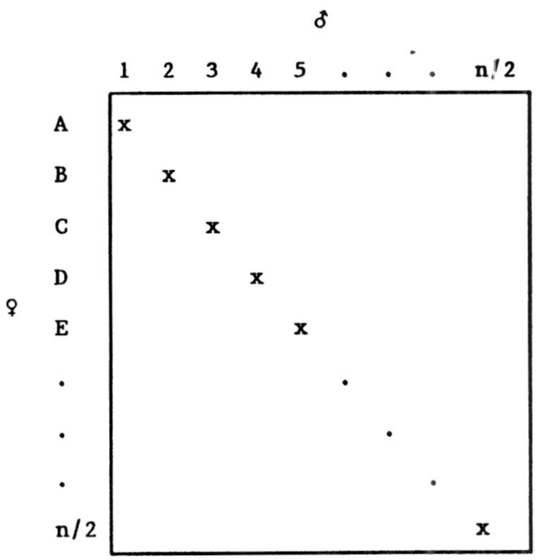

Single-pair mating design

Gain estimates: Provides good estimates of realized gains from full-sib family selection only. Expected gains based on estimates of general combining ability are not available for each parent. However, if there is evidence to suggest that non-additive genetic variances are negligible, gains expected from additive genetic variation may be estimated.

Parameter estimation: The component of variance or covariance among full-sib family means can be used to estimate 1/2 of the additive genetic variance or covariance only if non- additive genetic variance and covariance is known to be negligible.

Base population: Full pedigrees may be kept since controlled pollinations are made. If parents of unknown breeding value are mated to provide a base population for the next cycle of selection, mating each parent only once means that there is a strong probability that it will have been mated with an average or poorer parent. Thus, it may not be possible to select progenies only from parents with good breeding values, and gains will be reduced. But, because so few crosses are required, and because parent trees developing only one sex can be included, this type of mating design promotes maintenance of a large base population.

Cost: This mating design is the least complex mating scheme that permits maintaining complete pedigrees, and cost per parent is less than half that of a polycross.

3.2.4 Nested or hierarchical matings

Nested or hierarchical matings are those that mate groups of parents of one sex to different parents of the other sex, e.g. each of a group of males mated to a different subset of females. This design is useful primarily for estimating population parameters.

Progeny testing: Provides good estimates of gca effects for the rarer sex (males in the example above) if they are mated to a sufficient sample of the more numerous sex (females in the example above). Gca and sca effects are confounded in the more numerous sex since each is mated only once.

♀ \ ♂	1	2	3
A	x		
B	x		
C	x		
D		x	
E		x	
F		x	
G			x
H			x
I			x

Nested mating design

Gain estimates: Provides good estimates of realized gains that arise from additive genetic variation only for the rarer sex. Realized gains only from sca effects are reliable for parents of the more numerous sex unless there is evidence to suggest that non-additive gene actions are negligible. Expected gains that arise from either additive or dominance variance can be reliably estimated, but those from epistatic gene action are still poorly estimated.

Parameter estimation: The component of variance or covariance among parental groups of the rarer sex (males in the example above) estimates 1/4 of the additive genetic variance or covariance. The component of variance among parental groups of the more numerous sex estimates 1/4 of the additive genetic variance or covariance and 1/4 of the dominance genetic variance or covariance (under the assumption of no epistasis).

Base population: Full pedigrees may be kept since controlled pollinations are made. Parents of the more numerous sex will have been mated only once, therefore there is a strong probability that it will have been mated with an average or poorer parent of the rarer sex. Thus, it may not be possible to select progenies only from parents with good breeding values, and gains will be reduced. Furthermore, if selection is to be done among groups of the rarer sex based on their breeding values, many groups must be mated in order to provide a large selection differential. This mating design provides base populations with limited utility for selection.

Cost: This mating design is the least complex controlled-cross mating scheme with multiple matings per entry.

3.2.5 Factorial matings

Factorial mating designs require mating each member of a group of one sex with each member of a group of the other sex.

Progeny testing: Provides for good estimates of gca and sca effects.

Gain estimation: Provides for good estimates of realized gains and expected gains that arise from additive and non-additive genetic variation.

Parameter estimates: The component of variance or covariance among male and female groups estimates 1/4 of the additive genetic variance or covariance and the component of variance or covariance for the in-

♀ \ ♂	1	2	3	4	5	6	7	8	9
A	x	x	x	x	x	x	x	x	x
B	x	x	x	x	x	x	x	x	x
C	x	x	x	x	x	x	x	x	x
D	x	x	x	x	x	x	x	x	x
E	x	x	x	x	x	x	x	x	x
F	x	x	x	x	x	x	x	x	x
G	x	x	x	x	x	x	x	x	x
H	x	x	x	x	x	x	x	x	x
I	x	x	x	x	x	x	x	x	x
J	x	x	x	x	x	x	x	x	x

Factorial mating design

teraction of males and females estimates 1/4 of the dominance genetic variance or covariance (under the assumption of no epistasis).

Base population: Full pedigrees may be kept. The selection differential should not be limited since each entry is mated to half of the population. Thus, there is only a small probability that a parent with a good breeding value will have been mated only to parents with below-average breeding values.

Cost: The number of matings required for square factorials is $(N/2)^2$ where N is the total number of parents in the population. This number of matings becomes practically impossible for breeding populations with hundreds of entries.

3.2.6 Tester designs

Tester matings are factorials where a small number of parents of one sex are chosen for a particular reason. These testers may be chosen because they represent a particular genetic background in which a breeder wishes to evaluate other entries; or such pragmatic reasons as they begin to produce pollen earlier.

Progeny testing: Provides for good to excellent estimates of gca and sca effects if the number of tester parents is sufficient. Four to six tester parents are considered sufficient for loblolly pine (Zobel and Talbert 1984).

Gain estimation: Provides for good estimates of realized gains and expected gains that arise from additive and non-additive genetic variation. Breeding values for the tester parents will be more reliably estimated since they will be mated to more different genotypes than the non-tester parents.

Parameter estimates: The component of variance or covariance among male and female groups estimates 1/4 of the additive genetic variance or covariance and the component of variance or covariance for the interaction of males and females estimates 1/4 of the dominance genetic variance or covariance (under the assumption of no epistasis). The estimate of additive components from variances or co-variances

		♂				
		1	2	3	4	5
	A	x	x	x	x	x
	B	x	x	x	x	x
	C	x	x	x	x	x
	D	x	x	x	x	x
♀	E	x	x	x	x	x
	F	x	x	x	x	x
	.	.	.	.	.	.
	.	.	.	.	.	.
	.	.	.	.	.	.
	n	x	x	x	x	x

Tester mating design

among non-tester parents will be more accurate than estimates from tester parents because of their greater numbers.

Base population: Full pedigrees may be kept. The selection differential among the 4 to 6 tester parents will not be limited since each parent will be mated to a relatively large sample of parents. Thus, the probability that a parent with a good breeding value will have been mated only parents with below- average breeding values is quite small. The selection differential among non-tester parents, however, will be severely limited since each will have been mated to the same, small sample of parents and the total number of unrelated selections that can be made from tester matings cannot exceed the number of tester parents. Thus, breeders selecting from populations produced with tester mating designs must either accept small selection differentials on the majority of the population, or deal with increased co-ancestry.

Cost: The number of matings required for tester matings is m(N-m) matings where N is the total number of parents in the population and m is the number of tester parents. This mating design has been used to emphasize progeny testing and genetic parameter estimation (Zobel et al. 1972).

3.2.7 Full diallels

Full diallels require mating each entry with every entry, including reciprocals and selfs. This mating design provides the maximum amount of information on the population, but the inclusion of selfs makes both the execution and analysis difficult.

Progeny testing: Provides excellent estimates of both gca and sca effects as well as reciprocal and selfing effects.

Gain estimation: Provides for excellent estimates of realized gains and expected gains that arise from additive and non-additive genetic variation.

Parameter estimates: The component of variance or covariance among half-sib means (the component of variance for gca) estimates 1/4 of the additive genetic variance or covariance and the component of var-

♂

		1	2	3	4	5	6	7	8	9
	1	x	x	x	x	x	x	x	x	x
	2	x	x	x	x	x	x	x	x	x
	3	x	x	x	x	x	x	x	x	x
	4	x	x	x	x	x	x	x	x	x
♀	5	x	x	x	x	x	x	x	x	x
	6	x	x	x	x	x	x	x	x	x
	7	x	x	x	x	x	x	x	x	x
	8	x	x	x	x	x	x	x	x	x
	9	x	x	x	x	x	x	x	x	x

Full diallel mating design

iance or covariance for sca estimates 1/4 of the dominance genetic variance or covariance (under the assumption of no epistasis). Provides the maximum amount of information from a single mating design.

Base population: Provides the maximum opportunity to manage co-ancestry and maximize the selection differential since all possible matings were made.

Cost: Requires N^2 matings, where N is the number of parents in the population. The number of matings is practically impossible for breeding populations with hundreds of entries.

3.2.8 Modified half diallels

Modified half diallels are full diallels excluding reciprocals and selfs.

♀ \ ♂	1	2	3	4	5	6	7	8	9	10
1		x	x	x	x	x	x	x	x	x
2			x	x	x	x	x	x	x	x
3				x	x	x	x	x	x	x
4					x	x	x	x	x	x
5						x	x	x	x	x
6							x	x	x	x
7								x	x	x
8									x	x
9										x

Modified half diallel mating design

Progeny testing: Provides good estimates of both gca and sca effects, but no selfing or reciprocal effects.

Gain estimation: Provides for good estimates of realized gains and expected gains that arise from additive and non-additive genetic variation.

Parameter estimates: Provides the maximum amount of information on general and specific combining ability variances from a single mating design. The component of variance among half-sib means (the component of variance for gca) estimates 1/4 of the additive genetic variance or covariance and the component of variance for sca estimates 1/4 of the dominance genetic variance or covariance (under the assumption of no epistasis).

Base population: Provides the maximum opportunity to manage co-ancestry and maximize the selection differential since all possible matings were made.

Cost: Requires N(N-1)/2 matings, where N is the number of parents in the population. The number of matings is practically impossible for breeding populations with hundreds of entries.

3.2.9 Partial diallels

Partial diallels are systematic subsets of full diallels.

Progeny testing: Provides good estimates of both gca and sca effects if sufficient numbers of matings per entry are made.

Gain estimation: Provides for good estimates of realized gains and expected gains that arise from additive and non-additive genetic variation if sufficient numbers of matings per entry are made.

Parameter estimates: Provides good estimates of general and specific combining ability variances and co-variances if sufficient numbers of matings per entry are made. The component of variance or covariance among half-sib means (the components for gca) estimates

♀ \ ♂	1	2	3	4	5	6	.	.	.	n
1		x	x	x	x					
2			x	x	x	x				
3				x	x	x	x			
4					x	x	x	x		
5						x	x	x	x	
6							x	x	x	x
.	x							x	x	x
.	x	x							x	x
.	x	x	x							x
n	x	x	x	x						

One kind of a partial diallel mating design

1/4 of the additive genetic variance or covariance and the component of variance or covariance for sca estimates 1/4 of the dominance genetic variance or covariance (under the assumption of no epistasis).

Base population: Full pedigrees may be kept. The selection differential will be limited if each entry is not mated to a large a large sample of parents to ensure a high probability that a parent with a good breeding value will not have been mated only to parents with below-average breeding values.

Cost: The number of matings depends on the number of matings per entry.

3.3 Variations and Combinations of Mating Designs

Variations of these mating designs and combinations of designs have been used to address specific needs in tree breeding programs.

3.3.1 Positive assortative matings

Positive assortative mating (PAM) can be used with any of the general purpose mating designs that include controlled pollinations to increase expected genetic gains. PAM simply pairs the first- and second-ranked entries; the third- to fourth-ranked, etc.; until the two poorest entries are mated. PAM has been proposed to develop small populations of elite genotypes (Cotterill 1984) and to increase the frequency of extremely good genotypes for vegetative propagation (Foster 1986). PAM works increasingly well as heritabilities for the traits of interest increase, since good genotypes are more easily identified. For heritabilities of 0.3 or less, the additional expected gains and numbers of extreme genotypes are small (Mahalovich and Bridgwater 1989).

3.3.2 Disconnected mating designs

Disconnected mating designs have been used to subdivide large breeding populations into smaller subsets between which no matings take place. This reduces effort and costs and provides the possibility of sublining. Any of the designs above may be used in disconnected schemes, but disconnected factorials or modified half diallels are most commonly employed.

♂

		1	2	3	4	5	6	7	8	9
	A	x	x	x						
	B	x	x	x						
	C	x	x	x						
	D				x	x	x			
♀	E				x	x	x			
	F				x	x	x			
	G							x	x	x
	H							x	x	x
	I							x	x	x

Three disconnected, six-parent factorials

The attributes of these general purpose designs are the same as above, except that limitations on the realized selection differential will depend on the numbers of entries in each disconnected subset. Four x four or five x five factorials and six-parent modified half diallels are in common usage. With similar numbers of entries these two designs are similar in terms of information and providing a suitable base population. Peculiarities of flowering phenology of some species may favor disconnected factorials in terms of ease of making matings.

♀ \ ♂	2	3	4	5	6	8	9	10	11	12
1	x	x	x	x	x					
2		x	x	x	x					
3			x	x	x					
4				x	x					
5					x					
7						x	x	x	x	x
8							x	x	x	x
9								x	x	x
10									x	x
11										x

Two disconnected, six-parent modified half diallels

3.3.3 Complementary mating schemes

The use of multiple general purpose mating designs in the same population to better accomplish different specific goals has been termed ***complementary mating schemes*** (van Buijtenen et al. 1976). The most common combination is a relatively inexpensive method for estimating breeding values of parents and a controlled-cross design to produce a fully pedigreed base population for the next cycle of selection. For example, van Buijtenen and Lowe (1979) proposed the combination of polymix crosses and partial diallel mating schemes for a large loblolly pine breeding population.

3.3.4 Mating designs to manage co-ancestry and inbreeding depression.

Three fundamental approaches are currently in use or being considered for use, singly or in combination, to manage co-ancestry. These are regular systems of inbreeding, sublining, and adding a small population to the population hierarchy. A detailed discussion of these approaches is beyond the scope of this chapter. Thus, only a brief description of the systems will be given.

The potential for the use of regular systems of inbreeding in forest tree breeding programs has been summarized by Namkoong (1979). These schemes reach relatively high levels of inbreeding earlier than schemes that avoid inbreeding as long as possible. However, early avoidance schemes would accumulate inbreeding at greater rates than regular systems of inbreeding in the long term. The theoretical advantages of regular systems of inbreeding are not expected to occur for many generations, and the effects of levels of inbreeding less than selfing are not well known for most forest tree species. Thus, early avoidance of inbreeding followed by some regular mating scheme simply to effectively manage the breeding population may be the best approach for forest trees.

Sublining schemes subdivide large breeding populations into several smaller, closed populations. The subpopulations are closed in the sense that matings are made only within and not between subpopulations. Thus, inbreeding would occur within sublines, but could be avoided in propagules for plantation establishment simply by mating selected parents from different sublines (van Buijtenen and Lowe 1979).

Adding a small population to the population hierarchy accomplishes several things. First, a small population, say 50 or fewer, can be cycled faster than one with several hundred entries. Thus, gains can be realized sooner. Secondly, re-selection in an existing large, base population before mating and re-selection adds a cycle of selection and increases expected gains. Third, the larger base population can serve as a source of enrichment for the small population to manage co-ancestry and incorporate new, interesting phenotypes. Two methods of managing this new population have been proposed for forest trees. The concept of "nucleus" population breeding proposes a periodic two-way exchange of selected entries between a nuclear population and a larger base population (Cotterill 1989). "Elite" population breeding proposes a one-way transfer of entries from a base population to a smaller population of elite genotypes (Mahalovich and Bridgwater 1989). Both systems have the advantages mentioned above and both employ positive assortative mating to further increase expected gains from the small population. The "elite" population approach sacrifices some increment of potential genetic gain in the base population since genotypes selected in the elite population are not returned to the base population; but thereby avoids potential problems from relatedness between the base and elite populations.

Literature Cited

Bridgwater, F. E. and Franklin, E. C. 1985. Forest tree breeding: Strategies, achievements and constraints. In: Attributes of trees as crop plants. (M. G. R. Cannell and J. E. Jackson, eds.), 36-48. Institute of Terrestrial Ecology, Great Britain.

Burdon, R. D. and Shelbourne, C. J. A. 1971. Breeding populations for recurrent selection: conflicts and possible solutions. *N.Z.J. For. Sci.*, 1:174-193.

Byram, T. D., Lowe, W. J. and McKinley, C. R. 1987. Polymix crosses for rust resistance screening. In: Proc. of the 19th S. For. Tree Impr. Conf., College Station, TX, June 16-18, 39-44.

Cotterill, P. P. 1984. A plan for breeding radiata pine. *Silvae Genetica* 33:84-90.

Cotterill, P. P. 1989. The nucleus breeding system. In: Proc. of the 20th S. For. Tree Impr. Conf., Charleston, SC. June 26-30, 36-42.

Foster, G. S. 1986. Making clonal forestry pay: breeding and selection for extreme genotypes. In: Proc. IUFRO Conf. on Breeding Theory, Progeny Testing and Seed Orchards, Oct. 13-17, Williamsburg, VA, 582-590.

Mahalovich, M. F. and Bridgwater, F. E. 1989. Modeling elite populations and positive assortative mating in recurrent selection programs for general combining ability. In: 20th S. For Tree Impr. Conf., Charleston, SC, June 26-30, 43-49.

Namkoong, G. 1966. Inbreeding effects on estimation of genetic additive variance. *For. Sci.* 12: 8-13.

Namkoong, G. 1979. Introduction to quantitative genetics in forestry. USDA-Forest Service, Tech. Bull. No. 1588, 342 pp., Washington, D. C.

van Buijtenen, J. P. and Bridgwater, F. E. 1986. Mating and genetic test designs. Southern Cooperative Series Bull. No. 309, Louisiana St. Univ., 5-10.

van Buijtenen, J. P., Burley, J., Matheson, A. C., Stonecypher, R., and Thor, E. 1976. Mating Designs, In: Proc. of the IUFRO Joint Meeting of Genetic Working Parties on Advanced Generation Breeding. Bordeaux, 28-29.

van Buijtenen, J. P. and Lowe, W. J. 1979. The use of breeding groups in advanced generation breeding. In: Proc. of the 15th S. For. Tree Impr. Conf., Mississippi St. Univ, June 19- 21, 59-65.

Wiselogel, A. E. and van Buijtenen, J. P. 1988. Probability of equal mating in polymix pollinations of loblolly pine (Pinus taeda L.). *Silvae Genetica* 37(5-6): 184-187.

Zobel, B. J., Weir, R. J., and Jett, J. B. 1972. Breeding methods to produce progeny for advanced-generation selection and to evaluate parent trees. *Can. J. For. Res.*, 2: 399-345.

Zobel, B. J. and Talbert J. T. 1984. *Applied forest tree improvement.* Wiley and Sons, Inc., New York. 505 pp.

Chapter 4
Field Test Design

J. Loo-Dinkins[1]

Recognition of genetic value at the provenance, family or individual level is essential for tree improvement. The tree improvement worker can only see or measure the phenotype which expresses environmental effects as well as genetic ones. The greater the influence of the environment, the lower the heritability, and the phenotype becomes increasingly inaccurate as a measure of genetic value. Estimated means may be accurate in spite of low heritability, but only if the test has been well designed.

4.1 Importance of Field Test Design

This chapter describes field designs for genetic tests, and provides guidelines for choices, including discussions of plot layout and sample size. Considerations in matching common functions of genetic tests with appropriate designs are discussed. Finally, a brief overview of approaches available for improving the quality of data from test sites with high local variability is presented. A list of symbols is presented at the end of this chapter.

---

[1]*Forest Geneticist, Forestry Canada, Fredericton, New Brunswick, Canada*

L. Fins et al. (eds.), Handbook of Quantitative Forest Genetics, 96–139.

4.1.1 Sampling genes

Genetic field tests are established for a variety of purposes. In virtually all cases, however, a goal is to sample the genes in a given population and to draw conclusions about that population. The test materials may be provenances, with or without family structure, progeny groups with open-pollinated, half- or full-sib relationship, or clones, which may or may not represent progeny groups. The purposes of provenance testing are usually to evaluate sources for commercial seed collection or for selection of material for a breeding program, and sometimes to delineate seed transfer zones, or for elucidation of genetic architecture. Progeny testing originated as a procedure for evaluating parents in a breeding population or, in the forestry case, a seed orchard. Other common purposes of progeny testing are to provide a source for selection of the next generation of parents and to provide material for estimation of genetic parameters. Often in practice, genetic tests are established for all three purposes. Occasionally, these tests are converted into seed orchards. Clonal material is used to test ortets in hedging orchards, and may be used in trials to allow more accurate estimation of genetic parameters than is possible with sexually propagated material (Foster and Shaw 1988).

Certain basic principles of design apply for all testing purposes, but to obtain high quality information, each testing purpose has specific design requirements. Compromises are inevitable in multi-function tests and practical considerations include weighing the loss of potential gain against the cost of the ideal program.

4.1.2 Sampling environments

Adaptation to environmental conditions under which a population has developed may constitute an important component of overall performance. When trees are grown in a new environment, performance may change in more or less unpredictable ways. Thus, in addition to sampling genes, it is usually important to sample the range of commercial planting environments for which improved seed is destined.

Forest planting sites are often highly variable, and because of the size and purposes of tests, much of the variability is often unavoidable in test plantations. The test designs commonly used in forestry were developed in agriculture, but forestry sites are typically much larger and much more variable than agricultural sites. In general, the breeder expects the experimental design to reconcile two conflicting objectives: to test materials on environmentally variable sites (to represent commercial planting locations), and to allow efficient separation of genetic from environmental effects when evaluating the entries.

4.1.3 Consequences of mistakes

In general, poor experimental design leads to inaccurate estimates of parental and individual breeding values and imprecise genetic parameter estimates. These poor estimates lead to mistakes in selection and in estimating genetic gains. Assuming a closed population, a given degree of selection inaccuracy compounds over generations. That is, potential gain missed in any one generation reduces the potential throughout succeeding generations. Faulty parameter estimates may lead to poor choices in advanced generation breeding strategies. There is also a danger that breeding programs may be considered poor economic risks and be discontinued if potential gain is missed due to poor design.

4.2 What is a Field Test Design?

The arrangement of test materials in a field test constitutes an important part of the design, including the statistical experimental design as well as plot size and shape. The number of genetic entries and allocation of these to blocks and plots strongly influence the effectiveness of the experimental design. In addition, filler trees, check lots and border rows may be included in the field test design.

4.2.1 Arrangement of test materials

The simplest experimental design is the completely randomized design, which has the advantage, aside from being easiest to install, of having the maximum number of degrees of freedom for error, and thus the most sensitive F-tests for testing the treatment effects (Cochran and Cox 1957). In theory, the completely random design is appropriate if the test plantation is small and very uniform (Bridgwater and McKeand 1983). In practice, however, test plantations are rarely sufficiently uniform to obviate the need for blocking. Even when a site appears very uniform at planting, variability is likely to become more obvious over time, which could be partially accounted for by approximately square blocks (square is a logical shape in the case of no apparent environmental gradients). Patchy mortality, not an uncommon event in test plantations, may result in serious imbalance in a completely randomized design. Failure to block when it is warranted results in larger error sum of squares, thus making differences more difficult to detect.

A randomized complete block design, or a variation of it, is most commonly used for genetic testing of forest trees. Advantages include reduced impact of site variability due to blocking, the possibility of adjusting for block effects in selection and probable retention of balance or near-balance in the case of patchy mortality by dropping a block from the analysis. A randomized complete block design is usually appropriate unless the number of entries is so large that blocks will include an unacceptable amount of environmental variability.

Incomplete block designs allow a larger number of entries to be tested with small block size, thus reducing within-block variability. Layout, analysis and interpretation are more difficult, however, and comparisons among genetic entries across blocks are usually less precise than with a complete block design. Balanced incomplete block designs allow all planned comparisons to be made with the same degree of precision (Cochran and Cox 1957). A balanced lattice, for example, ensures that every pair of entries occurs together in a block, though the number of entries occurring in any given block is equal to the square root of the total number of entries (see Fig.4.1 for an example of a balanced lattice layout). This limits the total number of entries that can be tested to those with real square roots. Cheng (1986) outlined a method for constructing balanced incomplete block designs with nested rows and columns for reducing within-block heterogeneity in two directions. Friedman and Namkoong (1986) de-

Fig. 4.1. Example of a balanced lattice design with nine entries, four replicates and three blocks per replicate (from Cochran and Cox 1957).

BLOCK	REP 1	BLOCK	REP 2
1	1 2 3	1	1 4 7
2	4 5 6	2	2 5 8
3	7 8 9	3	3 6 9

BLOCK	REP 3	BLOCK	REP 4
1	1 5 9	1	1 6 8
2	2 6 7	2	2 4 9
3	3 4 8	3	3 5 7

scribed five unbalanced incomplete block designs for estimation of family means. Schutz and Cockerham (1966) discussed the efficiency of three incomplete block designs for selection. The applicability of an incomplete design for a given situation can be easily assessed if an appropriate data set exists as shown by Example 4.1.

The second consideration when arranging material in a plantation is the size and shape of plots. Single-tree plots randomized within blocks provide the most complete sampling of the within-site variation (Conkle 1963, Wright and Freeland 1960, Loo-Dinkins and Tauer 1987). The inevitable unplanned mortality always causes imbalance, however. Given the statistical tools available, such as PROC GLM of SAS® (SAS® Institute, Inc. 1985a), this is not usually a problem for a-

Example 4.1 Estimating efficiency of incomplete blocks

Suppose a breeder has collected data from a progeny test with a large number of entries arranged in complete blocks and would like to know whether smaller incomplete blocks would have been more efficient. This can be tested as follows (Ainsley et al. 1987):

1. Analyze data as a complete block design, following the original intent.
2. Estimate the error variance of family means as $\sigma^2_{bf}/b + \sigma^2_e/bn$, where σ^2_{bf} is the block by family interaction variance, σ^2_e is the error variance, b is the number of blocks, and n is the number of trees per family per site.
3. Superimpose incomplete sub-block boundaries on the data. It will not be possible to simulate a balanced design because the arrangement of families within blocks can obviously not be changed to accommodate the new design.
4. Reanalyze data as an incomplete block design. Incomplete block designs may be analyzed using the General Linear Models procedure of SAS® (SAS® Institute Inc. 1985a), but without recovery of interblock information. Giesbrecht (1986) explained a method for full analysis of incomplete block designs using SAS®. The interblock analysis is important because block totals contain information on the treatment differences as treatments are not all represented in all blocks (Cunningham and Henderson 1966).
5. Estimate the error variance of family means and compare with the complete block analysis. If the error variance is smaller with the incomplete block analysis, the incomplete block design would have been more efficient than the design used. This can provide valuable insight for design of future tests.

nalysis as long as the tools are applied correctly. When imbalance includes missing cells, as will be the case with single-tree plots, Type IV sums of squares should be used in a SAS® PROC GLM analysis. In cases of imbalance without missing cells, Type III or Type IV sums of squares may be used as they will be identical. Silvicultural thinning of single-tree plots is difficult, also inevitably resulting in imbalance. In addition, sibs cannot be easily compared in the field. Other considerations are addressed in section 4.3.1.

Multiple-tree plots, with members of a genetic entry planted in a row, square or rectangular shape, provide for easier establishment, maintenance and sib comparison (van Buijtenen 1983). In addition, analysis of plot means may be balanced, and will certainly be more balanced than with single-tree plots. Plot means analysis is less expensive in computer time than analysis based on individual trees. Environmental and genetic effects are likely to be confounded, however, because of incomplete sampling of within-site variability. The degree of confounding depends on the size of the plots, the number of blocks and the pattern of environmental variability. In general, the larger the plots, the more difficult it is to cleanly separate genetic from environmental effects.

A multiple-tree non-contiguous plot, consisting of plot members arranged at random within each block, as proposed by Libby and Cockerham (1980), provides almost as complete within-site sampling as single-tree plots (Lambeth et al. 1983). Analysis may be done on a plot means basis, and if interlocking sub-blocks are used, thinning is easy. Sib comparison is difficult however, and layout and tracking over time are more difficult than for the other designs. Interlocking sub-blocks, as depicted by Fig. 4.2, consist of complete sub-blocks superimposed on a given area comprising a conventional block (Libby and Cockerham 1980). With three interlocking blocks, a plantation can be thinned twice, with one-third of the original trees removed each time, without affecting balance and retaining equal spacing for all trees (with the exception of neighbors missing due to mortality). This design is especially useful for tests commonly called farm field tests, which are often initially planted at very close spacings in an attempt to simulate stand conditions at an early age. Nested designs, consisting of sets of families or provenances with family structure (compact family designs), can be considered to be a variation of randomized block designs. Sets or provenances, constituting the main plot, are randomized within blocks and family plots are randomized within main plots. This allows for ease of comparison within sets or provenances, and ensures that most competition will be intra-provenance, which is usually desirable. This design minimizes within-block variability for comparisons within sets or provenances (Bridgwater and McKeand 1983) which is appropriate if these are the comparisons of greatest interest. Comparisons among sets or provenances are usually imprecise, however, as main plots are usually large and are tested against an error with fewer degrees of freedom.

Fig. 4.2. Field layout for non-contiguous plots with interlocking sub-blocks (from Libby and Cockerham 1980). Each block consists of three sub-blocks, A, B and C in block 1 and D, E and F in block 2.

BLOCK 1	BLOCK 2
A B C A B C	D E F D E F
C A B C A B	F D E F D E
A B C A B C	D E F D E F
C A B C A B	F D E F D E
A B C A B C	D E F D E F
C A B C A B	F D E F D E
A B C A B C	D E F D E F
C A B C A B	F D E F D E
A B C A B C	D E F D E F
C A B C A B	F D E F D E

4.2.2 Allocation of test materials

The appropriate numbers of genetic entries to be tested (test sites, blocks and trees per plot) depend on the purpose of the test, financial constraints and environmental patterns of variability. For genetic parameter estimation, the number of entries must be sufficiently high to sample the range of genetic variability and to allow for precise estimation of variance components. Genetic correlation estimates are particularly prone to large sampling errors with low numbers of trees per entry (Cotterill and James 1984). A well-designed test with 50 or more entries (or crosses) should provide good estimates of genetic parameters (Namkoong and Roberds 1974).

The number of test sites required depends upon the existence or expectation of genotype-by-environment interaction and the type of tree desired. Usually trees that are broadly adapted to the range of environments within a breeding zone are desired for breeding populations, which necessitates sampling the range of variability. If trees specifically adapted to given environments are desired, the requisite number of test sites will be lower, with more trees per family planted at each site (van Buijtenen 1983). Lindgren (1985) pointed out that economic considerations may be as important in determining the appropriate number of test sites as biological ones. He developed a method for determining the optimum number of test sites based on the expected ratio of within-site to genotype by environment interaction variances and the marginal cost of adding an additional site. He determined that the optimum number of sites (s) and trees per family per site (n) maximize the following:

$$\frac{1}{\sqrt{\{1 + \sigma^2_{fe}/(s \cdot \sigma^2_f) + \sigma^2_e/(ns \cdot \sigma^2_f)\}}} \qquad \text{Eq. 4.1}$$

where σ^2_{fe} is genotype by environment interaction variance, σ^2_f is family variance, and σ^2_e is error variance. This is inversely proportional to the variance associated with a family mean. In general the larger this quantity, the more precise the test. Letting C=the number of units/entry that could theoretically be afforded if tested on zero sites (i.e. the number of seedlings per entry that could be produced to the point of out-planting) and A=the marginal cost of adding one site expressed as the number of trees/family sacrificed to add one more site, the total number of units per entry may be expressed as:

$$ns = C - As \qquad \textbf{Eq. 4.2}$$

Equation 1 is maximized when the following is minimized

$$\sigma^2_{fe}/s\sigma^2_f + \sigma^2_e/(C\text{-}As)\sigma^2_f \qquad \textbf{Eq. 4.3}$$

By multiplying both the numerator and denominator of the second term by σ^2_{fe}, letting $Q = \sigma^2_e/\sigma^2_{fe}$, and considering σ^2_{fe}/σ^2_f a constant, the quantity to be minimized becomes:

$$1/s + Q/(C\text{-}As) \qquad \textbf{Eq. 4.4}$$

The optimum t is found by differentiating the above equation with respect to t and equating the differential to 0. After some simplification, the optimum t may be expressed as:

$$s = C/[A * (1 + \sqrt{(Q/A)}\,)] \qquad \textbf{Eq. 4.5}$$

This approach is demonstrated by Example 4.2. If available, prior information about variances and the expected size of differences between means can be very useful in determining efficient numbers of blocks and trees per plot. A variety of approaches is available for determining the number of blocks or total number of trees per family required to detect significant differences between means. Cotterill and James (1984) discussed two methods for determining sample size: one based on size of detectable difference between families in stand-

Example 4.2 Estimation of optimum number of test sites

Suppose a breeder wishes to estimate the optimum number of test sites for a particular study, and can estimate from previous experiments that $Q = \sigma^2_e/\sigma^2_{fe}$ is approximately 5. If the breeder can afford to produce 200 seedlings per entry (C=200) to the point of out-planting, and has estimated that the marginal cost of adding one site would equal the savings realized by reducing the number of trees per entry by 20 (A=20), the theoretical optimum number of test sites may be estimated as follows (Lindgren 1985):

$$s = C/[A*(1+\surd(Q/A))]$$

$$Q = 5$$

$$C = 200$$

$$A = 20$$

$$s = 200/[20*(1+\surd(5/20))]$$

$$= 7 \text{ sites}$$

If genotype-by-environment interaction is known or expected to be larger, for example $Q = \sigma^2_e/\sigma^2_{fe} = 1$, the optimum number of sites would be estimated as:

$$s = 200/[20*(1+\surd(1/20))]$$

$$= 8 \text{ sites}$$

Note: This method has not been tested by the author of this chapter and numbers used in the example are not intended to reflect any real situation.

ard deviation units, demonstrated by Example 4.3, and the other based on the probability of detecting the best family. Their first approach is an adaptation of the sample-size equation given by Snedecor and Cochran (1980) for estimating sample size for comparison of two treatments (Equation 6.14.2). They estimated the minimum difference

Example 4.3 Estimating minimum number of trees per family required to detect differences between family means.

A breeder wishes to determine the appropriate number of trees per family to be tested (per site) and has some prior knowledge of variance structure in the population. By estimating the minimum difference in standard deviations between two family means that should be detected with given sample sizes, a sample size likely to result in significant differences may be determined as follows (Cotterill and James 1984):

Assume:

$\alpha = .05$

$p = .8$

$h^2 = .3$

$\beta = 2(1-.8) = .4$

$t = (1/4) * .3 = .075$

For n = 10:

$\delta/\sigma = (Z_{.05} + Z_{.4}) * \sqrt{\{(2 - 2(.075))/10\}} = (1.96 + .845) * .430$

$\delta/\sigma = 1.21$

For n = 20:

$\delta/\sigma = 2.805 * .304$

$= .853$

i.e. the difference between family means must be 1.21 standard deviations or greater to be detected for n=10 or .853 to be detected with n=20 trees per family. If the breeder has confidence in finding a difference of at least 1 standard deviation between families, trying n=15 would be a logical next choice. If the difference is expected to be less than .8, however, higher numbers of trees per family should be tested in the equation.

that should be detected between families in standard deviations (δ/σ) for a range of sample sizes (n), assuming single-tree plots as follows:

$$\delta/\sigma = (Z_\alpha + Z_\beta) \cdot \sqrt{\{)2 - 2t)n\}} \qquad \textbf{Eq. 4.6}$$

where Z_α denotes the two-tailed significance level, Z_β denotes the power of the test, ß = 2(1-p) where p = the desired probability of obtaining a significant result if the true difference is δ, t = the intraclass correlation (t = $h^2/4$ if half-sibs)[2] and n = the total number of trees per family. Cotterill and James' second approach is probably more useful because, as they pointed out, selection accuracy is of greater importance to the breeder than significant differences among means. They calculated the probability that a family with a true mean greater than that of a second family by a given amount will have a higher observed mean in a progeny test. Probabilities were tabulated over a range of standardized differences between true means and a range of numbers of individuals per family. Bridgwater et al. (1983) described a method for determining appropriate sample sizes based on coefficients of variation of entry means and observing how these change with varying numbers of locations, blocks, and trees per plot. Sanders (1989) explained a method for determining the power of a test, that is the probability of detecting significance if means are truly different, with varying numbers of blocks, using a function of SAS®. In some cases, efficiency of future testing may be improved by examining previously collected data to determine the minimum

---

[2] *Editor's note: With open-pollinated families, which often include full-sibs, selfs, and other progeny from related matings, t may be larger than $h^2/4$.*

number of blocks that would have been required to detect significant differences. In the case of a randomized complete block design with a random model the minimum number of blocks must satisfy:

$$nb\sigma^2_f/(\sigma^2_e+n\sigma^2_{bf}) > F_{(f-1),(f-1)(b-1),1-\alpha} - 1 \,, \qquad \textbf{Eq. 4.7}$$

where n = number of trees per plot, b = number of blocks, F = the tabulated F value (FTAB) for the given degrees of freedom and significance level (α) and f = number of families. Rearranging this expression, the minimum number of blocks (MINB) may be determined by solving for:

$$MINB = \{(\sigma^2_e+n\sigma^2_{bf})/n\sigma^2_f\} * (FTAB\text{-}1) \qquad \textbf{Eq. 4.8}$$

over a range of b.

An approximate minimum number of blocks required to attain significance can be determined as follows:

1. Obtain variance components σ^2_f, σ^2_{bf}, σ^2_e
2. Calculate the variance ratio, $VR=(\sigma^2_e+n\sigma^2_{bf})/n\sigma^2_f$.
3. Find tabulated F values (FTAB) for appropriate combinations of degrees of freedom and required significance level (α) for a range of b.
4. Calculate MINB=(FTAB-1) * VR for each b. The minimum number of blocks that would yield a significant F-test with the given variance structure is indicated when MINB = b.

This approach is demonstrated by Example 4.4. If a test consists of multiple-tree contiguous plots, the minimum number of blocks to give significant differences if the plots had consisted of single trees may be estimated by assuming that the error variance for single-tree plots, σ^2_{e1}, will approximately equal $\sigma^2_e + \sigma^2_{bf}$, estimated from the original data. VR becomes σ^2_{e1}/σ^2_f. The method should be applied with the goal of determining the relative efficiencies of established tests. The procedure does not give any indication of the probability of detecting significance if the experiment were to be repeated (resulting in different estimates of means and variances).

4.2.3 Other design considerations

Considerations when planning a field test, in addition to experimental design, include the use of filler trees, the number of border rows and use of check lots. Anomalous planting spots in a test plantation are often handled by planting filler trees that will not be measured, in an effort to reduce the impact of within-site variability. If the test is expected to be measured after crown closure, it is important that all planting spots be planted to provide a similar level of competition for all test trees. If the number of filler trees is large enough to increase the size of the plantation substantially, within-site variability is likely to be increased. In general, the larger the plantation, the higher the within-site variability. The author's rule of thumb is: if 10% or more of the planting spots require fillers, the site should not be considered for a test plantation.

One or preferably two border rows surrounding the entire plantation are necessary, especially if trees will be evaluated after the onset of

Example 4.4 Estimating the minimum number of blocks required to detect differences among family means

Suppose the analysis of a genetic field test containing six blocks yielded the variance components listed below. The breeder has a number of additional tests to establish under similar conditions and wishes to estimate the minimum number of blocks that would have allowed detection of significant differences among entries in this test.

Assume:

$f = 30$

$n = 5$

$\sigma^2_f = .01$

$\sigma^2_{bf} = .016$

$\sigma^2_e = .154$

$1\text{-}\alpha = .05$

b	FTAB-1	VR	MINB
2	.86	4.68	4.02
3	.65	4.68	3.04
4	.59	4.68	2.76
5	.55	4.68	2.57

The minimum number of blocks (MINB) that would have been required, given the above variance structure, is very close to 3. In practical application, however, caution would dictate rounding up to 4 blocks if this information were to be used in planning future tests.

competition. Libby and Cockerham (1980) suggested that combining the idea of border rows and demonstration plots can be useful, especially with single-tree or non-contiguous plot configuration in the test design.

For long-term growth and yield studies, with large family or provenance plots, one or two rows of border trees should surround each plot. Usually this means that the plots must contain many more trees than the minimum required for measurement, for example, with 8 x 8 tree plots, only 36 or 16 trees would be measured with one or two border rows, respectively. This necessitates a great deal of extra cost and effort but is done to avoid differential competition along plot edges which could have a serious effect on growth over the years (Bridgwater and McKeand 1983). Border trees should provide the same level of competition as the test trees and all plantation edges should be far enough from any forest edge to be beyond the range of influence.

Check lots are valuable for assessing genetic gain or for tying together disjoint series of progeny tests, but must be used with care. Check-lot material must be truly representative of the appropriate population for accurate gain estimation. The population that is appropriate depends upon what is being measured, for example, gain from the present cycle of selection or gain from all cycles of selection. There must be a sufficient number of plots of check material to allow precise comparisons, but, again, inclusion of check lots increases the size of the plantation, generally increasing the within-site variability.

4.3 Matching Design With Situation

Each testing function has an optimum experimental design. The design best for one purpose may be very poor for another. Thus, the testing objectives must be clear before the design is specified. The arrangement of blocks, and in some cases plot configuration, can be decided only after the test sites have been chosen. Gates (1983) recommended evaluating planting sites before the site preparation is done because local variability may be more obvious then than later. Careful evaluation of test sites before planting is crucial for effectively implementing a test design.

4.3.1 Function

The three most common purposes of genetic tests are estimation of genetic parameters, estimation of breeding values of parents (progeny tests) and selection of the next generation of parents. Genetic parameter estimation requires accurate and precise estimates of variance components for the population and environments of interest. Accuracy of variance component estimation depends largely upon the number of entries, while precision depends upon the number of trees per entry and the efficiency of within-site environmental sampling. The appropriate number of entries depends upon the heritability of the trait or traits of interest and on the mating design (Namkoong and Roberds 1974, Cotterill and James 1984). In a factorial mating design with at least six parents of one sex, as few as 50 crosses may suffice (Namkoong and Roberds 1974).

McCutchan et al. (1985) reported that when individual heritability is greater than .30, many small families are most efficient for parameter estimation. If heritability is less than .30, fewer large families were most efficient for a given plantation size. They also reported that efficiency of heritability estimation (measured by size of standard errors) could be improved by increasing the variance of family size, for a fixed total number of trees, when heritability was less than .30. This improved the efficiency of estimation because the positive effect of the larger-than-average families on the variance of the estimate outweighed the negative effect of the smaller-than-average families. McCutchan et al. (1989) concluded that it is more efficient to use larger-family-sized designs at the lower heritabilities than variable family size, however. In general, variance component estimation is considerably more difficult with unbalanced data (Searle 1979). Thus it seems reasonable to aim for balance when designing an experiment for variance component estimation.

The plot configurations that sample within-site variability most completely are the single-tree and non-contiguous plots (Wright and Freeland 1960, Conkle 1963, Libby and Cockerham 1980, Lambeth et al. 1983). If substantial mortality occurs, the single-tree plot design may also be analyzed as a non-contiguous plot design by combining two or more blocks, restoring some degree of balance at the plot level. A great advantage of this option is that blocks may be combined at the time of data analysis when mortality has occurred and is therefore known, and patterns of site variation may be evident, allowing more appropriate blocking.

Estimation of breeding values requires accurate and precise estimation of family means. White and Hodge (1988) examined the quality of selections made using Best Linear Prediction with differing numbers of test sites. In general, accuracy of breeding value estimation increases with higher numbers of test sites. For each site a sufficient number of individuals per family must be included to sample the genetic variability and within-site environmental variability adequately. Cotterill and James (1984) suggested that 10 to 20 individuals are sufficient for reliable estimation of family means in a well-designed test. The plot configuration should allow as complete as possible sampling of the within-site variability. In general, the single-tree plot arrangement is best for evaluation of families based on the aggregate performance of individual family members. For long-term studies of growth and yield under stand conditions, large square or rectangular plots may be necessary (Nance et al. 1986), though mixing genotypes in large plots may be more realistic, if commercial plantations will consist of bulk seed orchard seed. The precision of family means estimates in a randomized block design is expressed in terms of the error variance of a family mean as follows:

$$E_F = \sigma^2_{bf}/b + \sigma^2_e/nb. \quad \textbf{Eq. 4.9}$$

Table 4.1 presents results of a study comparing the effect of plot size and block number on the precision of family mean estimates in two loblolly pine progeny test sites (Loo-Dinkins and Tauer 1987). The efficiency of square plots, row plots, and non-contiguous plots is presented relative to that of single-tree plots. The procedure for such a comparison is fairly straightforward and could be applied to any test site for which data are available. The steps are:

1. Subtract family means from original data. If the data are unbalanced, a least-squares analysis such as the GLM procedure of SAS® will generate least-squares residual values (model : trait = family).
2. Superimpose block and plot boundaries on residual values to correspond to the designs of interest as shown by Example 4.5.
3. Analyze using the model:

$$\text{trait} = \text{block} + \text{plot(block)}$$

4. Estimate the error variance of a family mean:

$$E_F = \sigma^2_{bf}/b + \sigma^2_e/nb$$

Note that σ^2_{bf} = the variance among plots and can be considered a pure environmental variance. Estimation of family means is most precise for small E_F, resulting in better selections than when E_F is larger.

Accurate evaluation of individual genotypes is required for efficient within-family selection. The most important consideration is uniformity of environment within families. In addition, a sufficient number of trees per family must be tested to yield desired selection intensity. Assuming that family selection is done elsewhere, it is logical that the best design is large square or almost square family plots with no blocking. From the point of view of genetic efficiency,

Table 4.1. Relative efficiency of six plot configurations for estimation of family means on two loblolly pine progeny test sites with different patterns of environmental variability (Loo-Dinkins and Tauer 1987).

Plot Design	No. of Blocks	Patchy Variability				Gradient			
		σ^2_b	σ^2_{bf}	σ^2_e	Relative Efficiency[a]	σ^2_b	σ^2_{bf}	σ^2_e	Relative Efficiency[a]
25-tree squares	2	0.069	0.041	1.153	0.14	0.041	0.092	0.207	0.07
10-tree rows	5	0.064	0.018	0.160	0.48	0.081	0.012	0.240	0.50
5-tree rows	10	0.065	0.016	0.154	0.70	0.104	0.005	0.218	0.74
5-tree non-contiguous	10	0.066	0.000	0.173	0.94	0.104	0.004	0.219	0.75
2-tree rows	25	0.065	0.016	0.164	0.85	0.155	0.024	0.144	0.95
single-tree	50	0.063	0.167	-	1.00	0.138	0.182	-	1.00

[a] Relative efficiency = family mean error variance for the single-tree design plot design divided by the error variance for design i.

Example 4.5. Two alternative plot layouts for 15 trees per family at a single site.

Layout of hypothetical original plantation with 3 blocks and 5-tree row plots. Block boundaries are vertical, plot boundaries are horizontal.

Block 1	Block 2	Block 3
x x x x x	x x x x x	x x x x x
x x x x x	x x x x x	x x x x x
x x x x x	x x x x x	x x x x x
x x x x x	x x x x x	x x x x x
x x x x x	x x x x x	x x x x x
x x x x x	x x x x x	x x x x x
x x x x x	x x x x x	x x x x x
x x x x x	x x x x x	x x x x x

New block boundaries for single-tree plots, keeping the same number of families and the same block orientation.

Block

1	2	3	4	5	6	7	8	9	10	11	12	13	14	15
x	x	x	x	x	x	x	x	x	x	x	x	x	x	x
x	x	x	x	x	x	x	x	x	x	x	x	x	x	x
x	x	x	x	x	x	x	x	x	x	x	x	x	x	x
x	x	x	x	x	x	x	x	x	x	x	x	x	x	x
x	x	x	x	x	x	x	x	x	x	x	x	x	x	x
x	x	x	x	x	x	x	x	x	x	x	x	x	x	x
x	x	x	x	x	x	x	x	x	x	x	x	x	x	x
x	x	x	x	x	x	x	x	x	x	x	x	x	x	x

only one site is necessary or desirable, because trees cannot be easily compared across sites. A second site, providing backup, in case of unforeseen disaster destroying the first site is usually a good idea, however.

If one test is established for all three of the above purposes, compromises must be made. van Buijtenen (1983) suggested assigning priorities by deciding for a given program what the most important objectives are. For selection, the most appropriate design depends largely upon the anticipated gains to be made through family versus within-family selection. If heritability of the trait(s) of interest is low, greatest gains are made by selection at the family level (Falconer 1981), and the best design would consist of single-tree or non-contiguous plots with small block size. With moderate to high heritability, allowing greater gains by within-family selection, contiguous multi-tree plots may be better because of improved sib comparison.

4.3.2 Environment

Environmental variability can be roughly grouped into three categories; random, patchy and gradient. If variability were truly random, the choice of plot size and shape would be irrelevant, as a group of adjacent trees would sample as much of the within-site variability as would widely separated trees. This is rarely the case, however. A patchy type of variability results in environmental correlations between adjacent trees, with the strength of correlations likely to be roughly equal in all directions. When variability is patchy, multi-tree contiguous plots are likely to result in poor estimates of family means because of confounding of environmental and genetic effects at the plot level. A gradient results in environmental correlations between adjacent trees that are stronger in the direction perpendicular to the gradient than in other directions. A gradient can be easily handled if it is apparent when the trees are planted. As indicated by Table 4.1, short row plots may be as efficient or more efficient than non-contiguous plots of the same number of trees if they parallel the gradient, while blocks run perpendicular to the gradient. The most efficient design, however, for estimating genetic parameters or family means when a gradient is present, is single-tree plots in long narrow blocks perpendicular to the gradient. For within-family selection, sites with an obvious gradient should be avoided, or, the row plots should be planted perpendicular to the gradient, which is the exact opposite strategy as for between-family selection.

4.3.3 Practical considerations

In general, the more complicated designs (and the most efficient ones) are more difficult to establish and may involve higher costs than the usual multi-tree plots in randomized complete block designs. Mistakes in layout and tracking are more probable with non-contiguous than with contiguous plots, although this can be alleviated somewhat by use of computer-generated tally sheets. The cost and effort of establishment of new tests may be inversely proportional to resources available for maintenance and measurement of existing tests. Conversely, as the statistical efficiency of the design improves, the number of trees per entry to be tested on a given site may be reduced. For long-term experiments that will be measured repeatedly, the additional cost, if any, associated with initial layout is likely to be recovered manyfold in lower maintenance and measurement costs, and/or better data per dollar invested.

4.4 *A Posteriori* Data Adjustment For Local Variation

In many cases, breeders do not have the option of designing all of their own tests. Plantations established 15 to 20 years ago often must provide the basis for today's breeding activities, regardless of design problems that may be apparent in hindsight. It may be possible to improve the quality of data, in some cases, by removing a source of variability that is not accounted for in traditional analyses.

4.4.1 Definition of the problem

If a test plantation is established on a site having a fertility or moisture gradient, or having a patchy pattern of variability, adjacent trees will be more alike, on average, than trees farther apart. If plots are multi-tree and contiguous, there may be substantial environmental covariance confounded with genetic covariances in a given plot. This usually results in large plot (entry-by-block interaction) variance, which may substantially reduce the precision of a field test. Smith (1938) formulated an "empirical law of soil heterogeneity" based on the assumption that strength of environmental correlation is a function of distance between two points. Using data from uniformity trials for a number of agronomic crops, Smith determined that:

$$\log(Vx) = \log(V1) - b^{*}\log(x) \qquad \textbf{Eq. 4.10}$$

where Vx = the plot variance for plot size x, V1 = the variance associated with a plot of size 1 unit. The expression may be rearranged as:

$$Vx = V1 / x^b \qquad \textbf{Eq. 4.11}$$

The regression coefficient, b, ranges from 0 to 1, and the strength of environmental correlations between adjacent trees decreases with increasing b. The expression seems to be of limited use in forestry tests, as the relationship loses validity with gradient-based variability (Modjeska and Rawlings 1983, Magnussen 1989) and does not take into account plot shape or tree to tree variability (Correll and Cellier 1987). However, Smith's model has served as a launching point for a number of attempts to quantify a neighborhood effect in agriculture.

A data set can be quickly and easily examined for the presence and approximate size of average environmental correlations between neighboring trees using PROC CORR of SAS®. X and y coordinates must be assigned to each tree and any irregular edges must be padded with blanks to give the plantation a square or rectangular shape. A SAS® program to estimate average correlation coefficients between all trees in a moving 25-tree neighborhood as depicted in Fig. 4.3, is given in Example 4.6. The neighborhood size can be changed by writing: "Do row = x-i to x+i; Do col = y-j to y+j;" where i and j may be any (small) number. For example, i=1 , j=1 gives a 3 x 3 tree neighborhood.

Fig. 4.3. Example of one 25-tree neighborhood as created by the SAS® program (Example 4.6). As the neighborhood "moves" over the plantation, each tree, with the exception of the two edge rows, will be included in 25 neighborhoods as tree 1 through 25 (edge trees will be included in somewhat less than 25 neighborhoods). Average correlations between all pairs of trees in the moving neighborhood may be estimated.

	Columns										
Rows	1	2	3	4	5	6	7	8	.	.	.	70
1	x^1	x^2	x^3	x^4	x^5	x	x	x				x
2	x^6	x^7	x^8	x^9	x^{10}	x	x	x				x
3	x^{11}	x^{12}	x^{13}	x^{14}	x^{15}	x	x	x				x
4	x^{16}	x^{17}	x^{18}	x^{19}	x^{20}	x	x	x				x
5	x^{21}	x^{22}	x^{23}	x^{24}	x^{25}	x	x	x				x
6	x	x	x	x	x	x	x	x				x
7	x	x	x	x	x	x	x	x				x
8	x	x	x	x	x	x	x	x				x
.									.			.
.										.		.
.											.	.
50	x	x	x	x	x	x	x	x	.	.	.	x

4.4.2 Approaches for data adjustment

Wright (1978) proposed a method for improving the quality of information from a randomized complete block design by adjusting plot means on the basis of neighboring plot performance. He subtracted the average deviations from the family means of the two plots

Example 4.6 SAS® program to estimate correlations among neighboring trees (SAS® Institute Inc. 1985b).

```
*Create a temporary SAS® data set (temp) from an ASCI data file (dsname)
*with coordinate variables (x and y);
data temp; infile dsname; input x y trait;
proc sort data=temp; by x y;
*Create a new temporary SAS® data set (naybors) to set up 5 x 5 tree
*neighborhoods and assign trees position numbers 1 to 25;
data naybors; set temp; posn=0;
   do row=x-2 to x+2; do col=y-2 to y+2;
   posn=posn+1; output; end; end;
proc sort data=naybors; by row col posn;
*Transpose the data set to make the 25 position numbers variable names;
proc transpose data=naybors out=naybors prefix=posn;
   by row col; var trait;
*Estimate simple correlations between pairs of trees within neighborhoods *over
the entire plantation;
proc corr data=naybors; var posn1-posn25;
```

Output will include a matrix of average correlations between measurements for pairs of position numbers. The correlation coefficient for posn1 with posn2, for example, measures the average correlation between all adjacent trees in this orientation in the plantation. If environmental correlations are zero, i.e. microsite variability is random, estimates will vary around zero. If non-random (patchy or gradient) variability exists, correlations will be positive and highest between adjacent positions. If there is an environmental gradient, correlations will be higher in one direction than the other.

on each side from the subject plot mean. The method may be expressed as:

$$AP_i = P_i - [(P_{i-1} - F_{i-1}) + (P_{i-2} - F_{i-2}) + (P_{i+1} - F_{i+1}) + (P_{i+2} - F_{i+2})]/4 \qquad \textbf{Eq. 4.12}$$

where AP_i = the ith adjusted plot mean, P_i = the ith raw plot mean, P_{i-1}, P_{i+1}, P_{i-2}, P_{i+2} = plot means adjacent to and two plots away from plot i and F_{i-1} = the family mean of plot i-1. He analyzed the adjusted plot means using analysis of variance. Wright noted that this procedure could be considered equivalent to calculating plot means as a deviation from the regression line expressing site quality. He did not include blocks in his analysis because the block effect disappears with the adjustment, but he subtracted one degree of freedom for the adjustment. Analysis should always initially include the block effect, however, to ensure that neighbor analysis will never be worse than the traditional analysis.

A number of nearest-neighbor adjustment methods have been formulated for agricultural field tests such as variety trials. The methods are based on various smooth trend plus independent error models (also called "errors-in-variables" models) and they differ from each other chiefly in the estimation of the trend effect. The first such model was formulated by Papadakis in 1937 but only recently rediscovered (Bartlett 1978). Papadakis advocated adjustment of plot means by covarying their values on neighboring plot deviations from treatment means. Adjusted values are obtained as follows:

$$AP_i = P_i - \beta[(P_{i-1} - T_{i-1}) + (P_{i+1} - T_{i+1})]/2. \qquad \textbf{Eq. 4.13}$$

Terms are defined as above for Wright's work and T_{i-1} = the treatment mean for plot i-1. Papadakis' model may be applied iteratively, that is, residuals obtained from the analysis may be submitted to a second analysis, using the same model. Wilkinson *et al.* (1983) cautioned, however, that iteration may result in serious upward bias of treatment effects.

The moving block nearest-neighbor method proposed by Wilkinson *et al.* (1983) is formulated in a similar way to Papadakis' model except that no adjustment is made for treatment effects. The method may be expressed as:

$$AP_i = P_i - b * P_{Ni} \qquad \textbf{Eq. 4.14}$$

where P_{Ni} = the mean of neighboring plots. The real difference between the two methods is in the way in which the regression coefficient is estimated. Wilkinson's b is estimated as:

$$b = Cov(P_i, P_{Ni}) \ / \ Var(P_{Ni}) \qquad \textbf{Eq. 4.15}$$

Wilkinson *et al.* (1983) argued that their adjustment is better than Papadakis' because of biases introduced by subtracting the treatment effect and by iteration.

The first difference with errors-in-variables model discussed by Besag and Kempton (1986), provides a method for explicitly estimating the trend effect as follows:

$$P_i - P_{i-1} = T_i - T_{i-1} + u_i - u_{i-1} + w_i \qquad \textbf{Eq. 4.16}$$

where u_i, u_{i-1} represent the trend effects for plots i and i-1, respectively, and w_i is random error, commonly termed "white noise." The parameters are estimated using maximum likelihood. The differences $u_i - u_{i-1}$ are assumed to be random variables, uncorrelated with each other and with w_i.

The least-squares smoothing method proposed by Green *et al.* (1985) estimates a smooth curve for the trend effect operating in blocks of three adjacent plots. They assume that the trend effect should be locally linear and choose a smoothing constant to ensure a given degree of smoothness. The aim of least-squares smoothing is to minimize the average of:

$$\delta(u_{i-1} - 2u_i + u_{i+1})^2 + w_i^2 \qquad \textbf{Eq. 4.17}$$

where δ = a smoothing constant and other terms are defined above. The chief difficulty is in choosing an appropriate smoothing constant.

Bongarten and Dowd (1984) proposed an approach for removing trend effects in forest genetics tests using regression and spline techniques. Their method results in a smooth surface fitted through the data by modeling segments of rows and columns with cubic polynomial regressions using least squares methods, and fitting the segments end to end (constituting splines) subject to smoothing constraints. As in the least-squares smoothing technique (Green *et al.* 1985), all variation including the random error or white noise and family effects could be accounted for by the model if segments are short enough. Bongarten and Dowd (1984) suggested using segment lengths that reduce the residual error without reducing the family variance.

4.4.3 Complicating factors in forestry tests

Forestry sites, whether test or commercial plantations, are always heterogeneous, but unlike most agricultural fields, the primary source of variability may not be associated with an environmental gradient. Much of the heterogeneity often appears as patches and patch size may be so small that most of the variability is on a tree to tree basis. All of the a posteriori data adjustment approaches are most effective on gradients. If tree-to-tree variability is so great that average environmental correlations estimated as shown in example 4.6, between adjacent trees are close to zero ($r<.15$)[3], no data adjustment technique is likely to improve the situation. Thus the probability of success for a nearest neighbor or other data adjustment technique could be expected to be lower than for agricultural tests.

All formalized neighbor analysis approaches operate on a plot-means basis. This is more reasonable in agricultural than forestry tests, because plot sizes are typically much smaller and individual plants are rarely of interest in the former case. Individual trees are often of interest in forest genetic tests, however, and "patch size" may be little larger than plot size. A fully specified model on an individual tree

---

[3]*Loo-Dinkins, J.A. and Woods, J.H. Neighborhood data adjustment for local environmental variability in progeny tests: a case study with coastal Douglas-fir. Can. J. For. Res. Submitted.*

basis would require more memory than is often available, even in main-frame computers, as a neighborhood effect would be estimated for every tree. Thus adjustment on an individual tree basis is desirable, but estimation of trend effects is not practical.

Perhaps the most serious complication in forestry tests is competition. Trees go through a number of growth phases throughout a rotation, but from the point of view of correlation structure among neighbors, over the life of a genetic test, three phases must be considered. Before crown closure, when environment is primarily influencing growth independently of effects caused by the plantation, average correlations between neighboring trees are likely to be positive. After crown closure, competition may influence growth rates to the extent that correlations between neighbors are negative on average (Tuskan and McKinley 1984). There must be an unspecified period of time during which average correlations are close to zero. Thus meaningful adjustment for within-site variability must be carried out prior to crown closure, perhaps before selection age, to avoid confounding effects with competition. Approaches similar to those discussed above may be used to adjust for competition effects (Besag and Kempton 1986, Magnussen and Yeatman 1987). Adjusting out competition effects may not be so universally desirable as adjusting for environmental variability, however. It is interesting to note that adjusting for site heterogeneity is likely to increase the heritability, while adjusting for competition may decrease heritability.

4.4.4 Evaluation of adjustment methods

Opinions of proponents of nearest-neighbor analyses in the agricultural literature fall into two camps: that a posteriori data adjustment may be useful for salvaging a less than ideally designed experiment (Binns 1987, Besag and Kempton 1986, Bartlett 1978), and that the techniques have sufficient power to merit designing experiments with neighbor analysis in mind (Williams 1985, Wilkinson et al. 1983). Green et al. (1985) noted, after comparing least squares smoothing results with standard analysis of variance with randomized complete blocks, that the amount of improvement to be achieved by neighbor methods is small compared to what has already been achieved with randomized blocks. In forest genetics field tests, improvements achieved by neighbor methods are likely to be less impressive than for agricultural tests for reasons discussed above.

A neighbor method may remove environmental covariances in a field test if the heterogeneity is primarily caused by a gradient or by patches that are smaller than the blocks but larger than plots; that is if positive environmental correlations exist between neighboring trees with $r>.15$. If adjustment at the individual-tree level is desired, statistical models using a least squares method can probably not be fitted to estimate trend effects because of the size of the problem. If individual tree data are adjusted without a formalized model, prior adjustment for family effects is important to avoid removing much of the genetic variation. Adjusting for family effects prior to neighbor analysis may artificially inflate heritability, however, in a test with only two or three blocks. This occurs because a portion of the environmental covariance may be confounded with the covariance among family members, as would be the case if a family occurred on better-than-average microsites in each block. The higher the number of blocks, the lower the probability of such confounding.

Usual evaluation measures for data adjustment methods include the standard error of a difference between means, the error variance of family means, estimated heritabilities with and without adjustment and number of rank changes among families. None of these provide a direct answer to the real question; that is "is selection accuracy improved by data adjustment?". The only way to address this question in the short term is by simulation modeling. The validity of the results is limited by the set of assumptions upon which the model is based, as is always the case with simulation modeling. In general, data adjustment for the removal of microsite effects should not be attempted in the absence of evidence of environmental correlations between test trees, and in the author's opinion, should be used with the aim of improving selection rather than improving estimation of variance components.

Symbols Used in this Chapter

σ	=	standard deviation
σ^2_{bf}	=	block by family interaction variance or plot variance
σ^2_e	=	error variance
σ^2_f	=	family variance
σ^2_{fe}	=	genotype by environment interaction variance
h^2	=	heritability
b	=	number of blocks
f	=	number of families
n	=	number of trees per family per site
s	=	number of test sites
t	=	intraclass correlation
FTAB	=	tabulated F-value
MINB	=	minimum number of blocks
Z_α	=	two-tailed significance level
Z_β	=	power of the test
β	=	2(1-p)
p	=	desired probability of obtaining significant result if the true difference is δ
α	=	significance level
VR	=	variance ratio = $(\sigma^2_e + n\sigma^2_{bf})/n\sigma^2_f$
V_x	=	plot variance for plot size x
V1	=	variance associated with a plot of size 1 unit
P_i	=	i^{th} raw plot mean
AP_i	=	i^{th} adjusted plot mean
P_{Ni}	=	mean of neighboring plots
F_{i-1}	=	family mean of plot i-1
T_{i-1}	=	treatment mean for plot i-1
u_i	=	trend effects for plot i
w_i	=	random error
δ	=	smoothing constant

Literature Cited

Ainsley, A.E., Paterson, L.J., and Patterson, H.D. 1987. A method for predicting the efficiency of incomplete-block trials. *Biometrics* 43:55-59.

Bartlett, M.S. 1978. Nearest neighbor models in the analysis of field experiments. J. R. Statistical Soc. B 40:147-174.

Besag, J., and Kempton, R. 1986. Statistical analysis of field experiments using neighbouring plots. *Biometrics* 42:231-251.

Binns, M.R. 1987. Practical use of neighbour methods and alternatives in the analysis of field trials. *Can. J. Plant Sci.* 67:477-489.

Bongarten, B.C., and Dowd, J.F. 1984. "Regression and Spline Methods for Removing Environmental Variance in Progeny Tests". *19th South. For. Tree Impr. Conf.* College Station, TX. pp. 312- 319.

Bridgwater, F.E., and S.E. McKeand. 1983. "Experimental Design and How it Relates to Test Objectives and Sites". Proceedings of Servicewide Genetics Workshop. USDA Forest Service. Charleston, SC. pp. 103-117.

Bridgwater, F.E., Talbert, J.T., and Rockwood, D.L. 1983. "Field Design for Genetic Tests of Forest Trees". In *Progeny Testing of Forest Trees South. Coop. Series Bull.* No.275. Auburn, AL. pp. 28-39

Cheng, C. 1986. A method for constructing balanced incomplete-block designs with nested rows and columns. *Biometrika* 73:675- 700.

Cochran, W.G, and Cox, G.M. 1957. *Experimental Designs* 2nd Ed. John Wiley & Sons. New York.

Conkle, M.T. 1963. "The Determination of Experimental Plot Size and Shape in Loblolly and Slash Pines" N.C. State Univ. Sch. For. Tech. Rep. No. 17.

Correll, R.L., and Cellier, K.M. 1987. Effects of plot size, block size and buffer rows on the precision of forestry trials. *Aust. For. Res.* 17:11-18.

Cotterill, P.P., and James, J.W. 1984. Number of offspring and plot sizes required for progeny testing. *Silvae Genet.* 33:203- 209.

Cunningham, E.P., and Henderson, C.R. 1966. Analytical techniques for incomplete block experiments. *Biometrics* 22:829-842.

Falconer. D.S. 1981. *Introduction to Quantitative Genetics* 2nd Ed. Longman, New York.

Foster, G.S., and Shaw, D.V. 1988. Using clonal replicates to explore genetic variation in a perennial plant species. *Theor. Appl. Genet.* 76:788-794.

Friedman, S., and Namkoong, G. 1986. "Estimating Family Means Using Unbalanced Incomplete Blocks." In *Breeding Theory, Progeny Testing and Seed Quality.* IUFRO Conf. Proc. Williamsburg, VA. pp.457-468.

Gates, J.E. 1983. "Site Selection, Site Preparation and Test Layout". Proc. Servicewide Genetics Workshop. USDA Forest Service. Charleston, SC. pp.363-368.

Giesbrecht, F.G. 1986. Analysis of data from incomplete block designs. Biometrics 42:437-448.

Green, P., Jennison, C., and Seheult, A. 1985. Analysis of field experiments by least squares smoothing. *J. R. Statist.* Soc. B 47:299-315.

Lambeth, C.C., Gladstone, W.T., and Stonecypher, R.W. 1983. Statistical efficiency of row and noncontiguous family plots in genetic tests of loblolly pine. *Silvae Genet.* 32:24-28.

Libby, W.J., and Cockerham, C.C. 1980. Random non-contiguous plots in interlocking field layouts. *Silvae Genet.* 29:183-190.

Lindgren, D. 1985. Cost-efficient number of test sites for ranking entries in field trials. *Biometrics* 41:887-893.

Loo-Dinkins, J.A., and Tauer, C.G. 1987. Statistical efficiency of six progeny test field designs on three loblolly pine (Pinus taeda L.) sites types. *Can. J. For. Res.* 17:1066-1070.

Magnussen, S. 1989. Interplant interactions and their influence on within and among plot variances. *Scand. J. For. Res.* 4:369- 377.

Magnussen, S., and Yeatman, C.W. 1987. Adjusting for inter-row competition in a jack pine provenance trial. *Silvae Genet.* 36:206-214.

McCutchan, B.G., Namkoong, G., and Geisbrecht, F.G. 1989. Design efficiencies with planned and unplanned unbalance for estimating heritability in forestry. *For. Sci.* 35:801-815.

McCutchan, B.G., Ou, J.X., and Namkoong, G. 1985. A comparison of planned unbalanced designs for estimating heritability in perennial tree crops. *Theor. Appl. Genet.* 71:536-544.

Modjeska, J.S, and Rawlings, J.O. 1983. Spatial correlation analysis of uniformity data. *Biometrics* 39:373-384.

Namkoong, G., and Roberds, J.H. 1974. Choosing mating designs to efficiently estimate genetic variance components for trees. *Silvae Genet.* 23:43-53.

Nance, W.L., McCutchan, B.G., Talbert, C.B., Buford, M.A., Foster, G.S.,and Sprinz, P. 1986. "Experimental Approaches for Evaluating Genetic Effects on Stand Growth and Yield" In *Statistical Considerations in Genetic Testing of Forest Trees* South. Coop. Series Bull. No. 324 Gainesville, FL. pp.22-39.

Sanders, W.L. 1989. Use of Proc GLM of SAS® (or a similar linear model computing tool) in research planning. *HortSci.* 24:40-45.

SAS® Institute Inc. 1985a. *SAS® Users Guide : Statistics* Version 5 Ed. Cary, NC.

SAS® Institute Inc. 1985b. *SAS® Users Guide : Basics* Version 5 Ed. Cary, NC.

Searle, S.R. 1979. *Notes on Variance Component Estimation : A Detailed Account of Maximum Likelihood and Kindred Methodology.* Cornell Univ., Ithaca, New York.

Shutz, W.M., and Cockerham, C.C. 1966. The effect of field blocking on gain from selection. *Biometrics* 22:843-863.

Smith, F.H. 1938. An empirical law describing heterogeneity in the yields of agricultural crops. *J. Agr. Sci.* 28:1-23.

Snedecor, G.W., and Cochran, W.G. 1980. *Statistical Methods* 7th Ed. The Iowa State Univ. Press. Ames.

Tuskan, G.A., and McKinley, C.R. 1984. The use of competition indices in advanced-generation selection. *Silvae Genet.* 33:209-215.

van Buijtenen, J.P. 1983. "Progeny Test Design and Analysis". 18th South. For. Tree Impr. Conf. Proc. Athens, GA. pp. 161-170.

White, T.L., and Hodge, G.R. 1988. Best linear prediction of breeding values in a forest tree improvement program. *Theor. Appl. Genet.* 76:719-727.

Wilkinson, G.N., Eckert, S.R., Hancock, T.W., and Mayo, O. 1983. Nearest neighbour (NN) analysis of field experiments. *J. R. Statist.* Soc. B 45:151-211.

Williams, E.R. 1985. A criterion for the construction of optimal neighbour designs. *J. R. Statist.* Soc. B 47:489-497.

Wright, J.W. 1978. An analysis method to improve statistical efficiency of a randomized complete block design. *Silvae Genet.* 27:12-14.

Wright, J.W., and Freeland, F.D. 1960. "Plot Size and Experimental Efficiency in Forest Genetic Research". Mich. State Univ. Agric. Exp. Stn. Tech. Bull. No. 280.

Chapter 5
Concepts of Selection and Gain Prediction

G.R. Hodge and T.L. White[1]

Selection is defined as the differential reproduction of genotypes, i.e., the relative number of gametes that parents contribute to the next generation depends in some way on their genotypes. In natural selection of forest trees, differential reproduction of genotypes is affected by traits like survival and adaptability. In artificial selection of forest trees, humans intervene in the reproductive process in order to increase gamete contribution from genotypes considered favorable. Successful artificial selection means that individuals with the most desirable genes contribute the most offspring in the next generation. In forestry, desirable trees are those that are, for example, disease resistant, faster growers or have higher specific gravity. Selection does not create new genes, but rather accumulates the best combinations of genes.

Within constraints on cost, population size and diversity, etc., the purpose of selection is always to maximize the genetic worth of the

[1]*Geneticist and Associate Professor, Department of Forestry, University of Florida, Gainesville, Florida*

L. Fins et al. (eds.), Handbook of Quantitative Forest Genetics, 140–194.

selected population; however, this task is made difficult by the fact that we cannot measure genotype (or genetic worth) of an individual directly. Instead, we are limited to measuring the phenotype (which is modified by the environment) of an individual or its relatives. Then, with a knowledge of genetic theory and certain genetic parameters, we use phenotypic measurements to predict the genetic worth of all candidates available for selection (these candidates may be individual trees, families, clones, etc.). Since the goal of making selections is to maximize genetic gain, it is then logical to select those candidates with the highest predicted genetic worth.

5.1 Chapter Organization

The purpose of this chapter is to illustrate an approach to selection and gain prediction that is applicable to situations ranging from mass selection of individuals to indirect selection for multiple traits using multiple measurements from diverse types of genetic tests. For all selection situations, the basic approach to making selections and predicting gain is the same:

1) predict genetic worth (g) of all candidates as a linear function of observed data (y),
2) rank and select candidates on the basis of their predicted genetic worth $\hat{g}$, and
3) calculate the predicted genetic gain ($\Delta\hat{g}$) as the average of the predicted worth of those selected minus the average genetic worth of all candidates.

We begin by deriving a general prediction formula applicable when predicting a single genetic value from a single phenotypic observation. Then for a number of selection scenarios (mass selection, family selection, indirect selection), we demonstrate how this formula reduces to some commonly used gain formulae, and illustrate the use of the formulae with numerical examples. The same basic approach is then developed to predict genetic values and make selections in more complex situations (e.g., using multiple and diverse phenotypic observations to predict a single genetic value).

This chapter assumes some familiarity with basic concepts in genetics such as heritability, family heritability, and genetic correlation. It also assumes an understanding of some statistical concepts and notation, in particular, summation notation, linear models, and variances and covariances of linear functions. For the reader who is unfamiliar with these statistical concepts, Neter and Wasserman (1974, Ch. 1) provide an excellent review.

5.2 Derivation of General Prediction Formula

The problem is to use some function of a phenotypic observation (i.e., a random variable y) to predict an unobservable genetic value (i.e., a random variable g). We assume that certain parameters associated with the random variables y and g are known, specifically the first moments (expected values, E(g) and E(y)), and second moments (Var(y), Var(g) and Cov(g,y)). In real life situations first and second moments are never really known, but we may be able to obtain good estimates of them. Then, within the class of linear functions, $\hat{g}$ = a

+ by, where $\hat{g}$ is the predicted genetic value, and the coefficients a and b are unknown and must be solved for. To solve for a and b such that $\hat{g}$ is a good predictor of the true genetic value g, first define a quantity Q such that $Q = E(g - \hat{g})^2$ and then solve for the coefficients a and b that minimize Q (i.e., that minimize the squared error of prediction). By expanding the squared term, substituting for $\hat{g}$ and distributing the expected value operator, Q is:

$$
\begin{aligned}
Q &= E(g - \hat{g})^2 \\
&= E(g^2 - 2g\hat{g} + \hat{g}^2) \\
&= E(g^2 - 2g(a+by) + (a+by)^2) \\
&= E(g^2 - 2ga - 2gby + a^2 + 2aby + b^2y^2) \\
&= E(g^2) - 2aE(g) - 2bE(gy) + a^2 + 2abE(y) + b^2E(y^2).
\end{aligned}
$$

To find the values of a and b that minimize Q, take the partial derivatives of Q with respect to both a and b, set the two equations equal to zero, and solve simultaneously.

The partial derivative of Q with respect to a is $\partial Q/\partial a = -2E(g) + 2bE(y) + 2a$. Setting $\partial Q/\partial a$ to zero, the solution for a is

$$
\begin{aligned}
0 &= -2E(g) + 2bE(y) + 2a \\
a &= E(g) - bE(y). \qquad \textbf{Eq. 5.1}
\end{aligned}
$$

The partial derivative of Q with respect to b is $\partial Q/\partial b = -2E(gy) + 2aE(y) + 2bE(y^2)$. By substituting the solution for a into this equation, $\partial Q/\partial b$ can be expressed as

$$\begin{aligned}\partial Q/\partial b &= -2E(gy) + 2[E(g)-bE(y)]\,E(y) + 2bE(y^2)\\ &= -2E(gy) + 2E(g)E(y) - 2bE(y)E(y) + 2bE(y^2).\end{aligned}$$

Setting $\partial Q/\partial b$ to zero, the solution for b is

$$0 = -2E(gy) + 2E(g)E(y) - 2bE(y)E(y) + 2bE(y^2)$$
$$2E(gy) - 2E(g)E(y) = -2bE(y)E(y) + 2bE(y^2)$$
$$E(gy) - E(g)E(y) = b[-E(y)E(y) + E(y^2)]$$
$$b[-E(y)E(y) + E(y^2)] = E(gy) - E(g)E(y)$$
$$b[E(y^2) - E(y)E(y)] = E(gy) - E(g)E(y)$$
$$b\ Var(y) = Cov(g,y)$$
$$b = Cov(g,y)/Var(y). \qquad \textbf{Eq. 5.2}$$

The coefficients a and b are functions of E(g), E(y), Var(y), and Cov(g,y) (Eqs. 5.1 and 5.2). The latter parameters are assumed known, and in practice may be estimated from available data. Then the genetic worth of any particular individual (g) can be predicted from an observed phenotypic observation (y) with the general formula

$$\begin{aligned}\hat{g} &= a + by\\ &= E(g) - bE(y) + by\\ &= \tau - b\alpha + by \qquad \textbf{Eq. 5.3}\\ &= \tau + b(y-\alpha)\\ &= \tau + [Cov(g,y)/Var(y)]\ (y-\alpha)\end{aligned}$$

where $\hat{g}$ = predicted genetic worth, τ = E(g) = average genetic worth for all candidates, Cov(g,y) is the covariance of the observed data with the genetic worth, Var(y) is the variance of the observed data, y is the observed data that pertains to the particular individual, and α = E(y) = expected value of all observations.

When all individuals for which genetic values are being predicted are from a single population, it is often convenient to assume that E(g) = τ = 0. Thus, all predicted genetic values will center around zero, and are expressed as deviations from the mean genetic value for the population. Thus $\hat{g}$ is predicted as

$$\begin{aligned} \hat{g} &= [\mathrm{Cov}(g,y)/\mathrm{Var}(y)]\ (y-\alpha) \\ &= b(y-\alpha). \end{aligned} \qquad \textbf{Eq. 5.4}$$

Since τ is a constant, rankings of individuals and genetic gain predictions based on Eqs. 5.3 and 5.4 will be identical if all candidates are from a single population and hence have equal τ values. Throughout the remainder of this chapter, genetic values will be predicted using Eq. 5.4. Genetic values predicted with Eqs. 5.3 and 5.4 are sometimes called best linear predictions (BLPs) because, among the class of linear functions of the data, the predictions are best under the criterion that the average squared error of prediction ($E(g-\hat{g})^2$) is minimized over repeated sampling. There are a number of other attractive properties associated with BLPs (Henderson 1973, 1977, 1984). These include: 1) correlation between true and predicted genetic value ($\mathrm{Corr}(g,\hat{g})$) is maximized; 2) the probability of a correct ranking of genotypes is maximized; and 3) selection done on the basis of predicted genetic values maximizes the average genetic value of the selected individuals. Properties 2 and 3 assume a joint probability distribution of g and y that is multivariate normal.

5.2.1 Precision of prediction

A useful measure of the precision of predicted genetic values is the correlation between the predicted and true genetic values ($\mathrm{Corr}(g,\hat{g})$).

Although the true genetic values are unknown and unobservable, $Corr(g,\hat{g})$ can be calculated if the first and second moments are known or estimated. The correlation between predicted and true genetic values is

$$Corr(g,\hat{g}) = Cov(g,\hat{g})/[Var(g)\ Var(\hat{g})]^{1/2}. \qquad \textbf{Eq. 5.5}$$

The only *known* term in Eq. 5.5 is Var(g), the genetic variance for the trait being predicted. To solve for $Corr(g,\hat{g})$, we need to express $Cov(g,\hat{g})$ and $Var(\hat{g})$ in terms of other *known* parameters. The variance of predicted genetic values ($Var(\hat{g})$) is

$$\begin{aligned} Var(\hat{g}) &= Var(a+by) \\ &= Var(by) \\ &= b^2 Var(y) \qquad \textbf{Eq. 5.6} \\ &= [Cov(g,y)/Var(y)]^2\ Var(y) \\ &= [Cov(g,y)]^2/Var(y). \end{aligned}$$

The covariance of predicted and true genetic values ($Cov(g,\hat{g})$) is

$$\begin{aligned} Cov(g,\hat{g}) &= Cov(g,\ a+by) \\ &= Cov(g,a) + Cov(g,by) \\ &= b\ Cov(g,y) \qquad \textbf{Eq. 5.7} \\ &= Cov(g,y)\ Cov(g,y)/Var(y) \\ &= Var(\hat{g}). \end{aligned}$$

Using Eq. 5.6 and Eq. 5.7 to substitute into Eq. 5.5, $Corr(g,\hat{g})$ is

$$\begin{aligned} Corr(g,\hat{g}) &= Var(\hat{g})/[Var(g)\ Var(\hat{g})]^{1/2} \qquad \textbf{Eq. 5.8} \\ &= [Var(\hat{g})/Var(g)]^{1/2}. \end{aligned}$$

$Corr(g,\hat{g})$ ranges from 0 to 1, and since Var(g) is a constant, $Var(\hat{g})$ ranges from 0 to Var(g). When the relationship between the observed y and the predictand g is strong, precision of prediction will be high, $Corr(g,\hat{g})$ will be near one, and $Var(\hat{g})$ will be near the genetic variance, Var(g). If the relationship between g and y is weak, both $Corr(g,\hat{g})$ and $Var(\hat{g})$ will be near zero.

5.3 Mass Selection

To apply the above formulae to the simplest case, consider mass selection, the selection of individual trees on the basis of their phenotypes alone. Mass selection can be for total genetic worth (as for vegetatively propagated material), or additive genetic worth (as for a seed production area or grafted seed orchard). Throughout this chapter, we consider only prediction of additive genetic worth, although the theory is completely general and applies to the prediction of other types of genetic values. A linear model for the phenotypic observations on individual trees is

$$y_i = \mu_y + g_i + e_i \qquad \textbf{Eq. 5.9}$$

where,

y_i = phenotypic observation on i^{th} tree, $E(y_i)=\mu_y$, $Var(y_i)=\sigma^2_y$,
g_i = additive genetic effect (also called breeding value) for the ith tree, $E(g_i)=0$, $Var(g_i)=\sigma^2_a$,
e_i = error for the ith tree, $E(e_i)=0$, $Var(e_i)=\sigma^2_e$, and there is no covariance between terms, i.e., $\sigma^2_y = \sigma^2_a + \sigma^2_e$.

We want to predict the breeding value (g) for trait y for each tree using Eq. 5.4, thus $\hat{g}$ = [Cov(g,y)/Var(y)] (y-α). Substituting for y using the linear model for y in Eq. 5.9, Cov(g,y) = Cov(g, μ + g + e) = Cov(g,g) = Var(g) = σ^2_a (all i subscripts have been deleted to simplify notation). Since Var(y) = σ^2_y, $\hat{g}$ can be expressed as

$$\begin{aligned}\hat{g} &= [\sigma^2_a/\sigma^2_y]\,(y-\mu_y)\\ &= h^2\,(y-\mu_y)\end{aligned} \qquad \textbf{Eq. 5.10}$$

where h^2 = narrow sense heritability = σ^2_a/σ^2_y, and μ_y = E(y) = α.

For the case of mass selection, the correlation of the true and predicted breeding value reduces to the square root of the heritability:

$$\begin{aligned}\mathrm{Corr}(g,\hat{g}) &= [\mathrm{Var}(\hat{g})/\mathrm{Var}(g)]^{1/2}\\ &= [\mathrm{Cov}(g,y)^2 \,/\, \{\ \mathrm{Var}(y)\ \mathrm{Var}(g)\ \}]^{1/2}\\ &= [\mathrm{Var}(g)^2 \,/\, \{\ \mathrm{Var}(y)\ \mathrm{Var}(g)\ \}]^{1/2}\\ &= [\mathrm{Var}(g)/\mathrm{Var}(y)]^{1/2}\\ &= [\sigma^2_a/\sigma^2_y]^{1/2}\\ &= h.\end{aligned} \qquad \textbf{Eq. 5.11}$$

5.3.1 Example 1: Prediction of breeding values from mass selection

To illustrate the application of the prediction formula (Eq. 5.4), consider hypothetical parameters for 10-year-old coastal Douglas-fir: E(y)=α=600 cm, σ^2_y = 14,138 cm^2 and h^2 = 0.157. The phenotypic standard deviation is σ_y = 119 cm, thus approximately 95% of the trees will be between 840 cm and 360 cm tall at age 10, assuming a normal distribution. Consider 5 trees with heights of 840, 720, 600, 480, and 360 cm, i.e., $\alpha+2\sigma_y$, $\alpha+\sigma_y$, α, $\alpha-\sigma_y$, $\alpha-2\sigma_y$, respectively. Then, assuming h^2 = 0.157 and using Eq. 5.10, $\hat{g}$ for the tallest tree is $\hat{g}$ =

0.157 * (840 - 600) cm = +37.68 cm. Predicted breeding values ($\hat{g}$) for the other four trees are +18.84, 0, -18.84, and -37.68 cm, respectively. Note that predicted breeding value is calculated by regressing the adjusted phenotypic record (y-α) towards zero, the assumed genetic average (in fact, this is the origin of the term linear regression). For example, the tree 240 cm above the mean is regressed back to a predicted genetic value that is 37.68 cm above the mean.

Phenotypic values above the mean are regressed down toward zero to predict the genetic value, and phenotypic values below the mean are regressed up toward zero to predict genetic value. The regression coefficient is the heritability, and the higher the heritability, the less the observed phenotypic deviation (y-α) is regressed toward zero when predicting the genetic value (Figure 5.1). Heritability measures the relative amount of genetic to total variance, and thus, in a sense, the relative amount of a phenotypic deviation (y-α) that we expect to be due to genetic superiority. For example, for the 840 cm tree, if h^2 = 0.157 then $\hat{g}$ = +37.68 cm, while if h^2 = 0.5 then $\hat{g}$ = +120 cm. Thus, only 15.7% of the 240 cm deviation from the phenotypic mean is 'credited' to genetic effects if h^2 = 0.157, while 50% of the deviation is 'credited' if h^2 = 0.5. Additionally, the higher the heritability, the more precise are the predicted genetic values, as reflected in the Corr(g,$\hat{g}$). From Eq. 5.11, if h^2 = 0.157 then Corr(g,$\hat{g}$) = 0.40, and if h^2 = 0.5 then Corr(g,$\hat{g}$) = 0.71.

Note that all individuals with the same phenotypic value (y) have the same predicted breeding value ($\hat{g}$). This does not necessarily mean

Selection and Gain

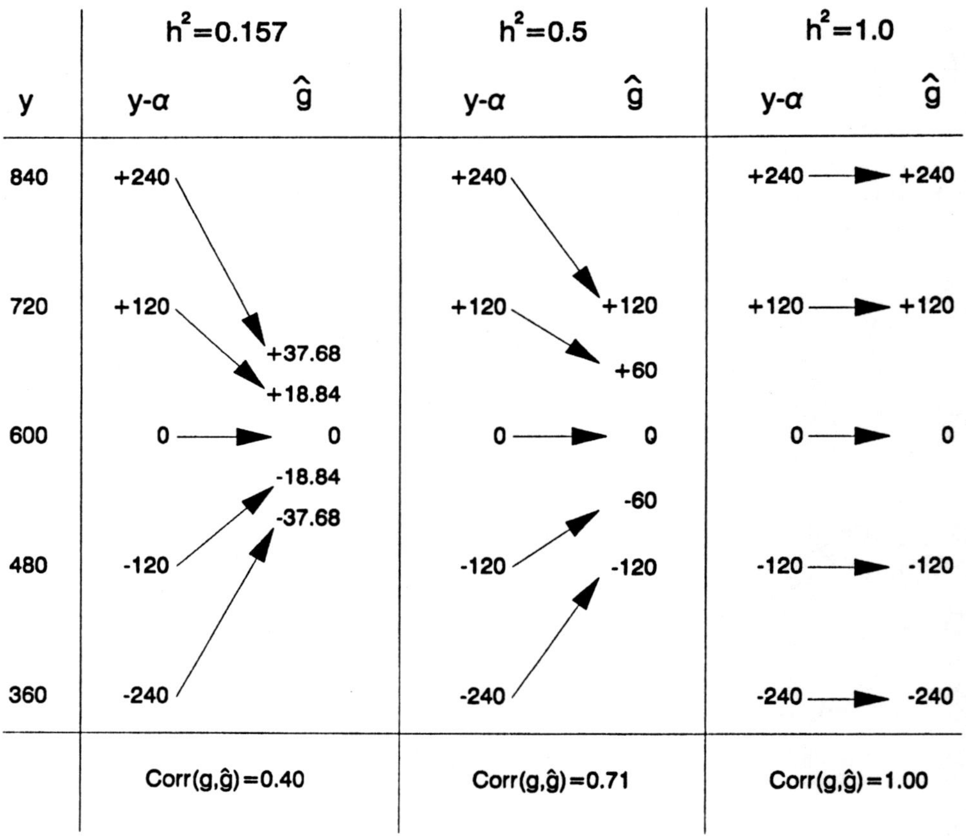

Figure 5.1 Breeding value is predicted by regressing the phenotypic deviations toward zero. When prediction is from a single record (as in mass selection) the regression factor is heritability. Adapted from Van Vleck et al. 1987.

they have the same true breeding value (g). For example, 15 trees, each 840 cm tall, would all have a predicted breeding value of +37.68 cm. The true breeding values will vary around this number, i.e., some of those trees would have a true value higher than +37.68 cm, and others lower. However, the best linear prediction that we can make for each individual tree is +37.68 cm.

5.3.2 Genetic gain from mass selection

After genetic values have been predicted for all individuals, selections can be made and an expected genetic gain calculated. Selection differential (S) is defined as the average phenotypic value of the selected individuals minus the overall phenotypic average, $S = (\bar{y}_S - \mu_y) = (\bar{y}_S - \alpha)$. Assume that N individuals are selected. Predicted genetic gain ($\Delta\hat{g}$) is equal to the average of predicted breeding values of the selected individuals.

$$\begin{aligned}
\Delta\hat{g} = \Sigma\hat{g}_i/N &= \Sigma[h^2\,(y_i-\mu_y)]/N \\
&= h^2\,\Sigma(y_i-\mu_y)/N \\
&= h^2\,(\Sigma y_i/N - \mu_y) \quad \{\Sigma y_i \text{ for those selected}\} \\
&= h^2\,(\bar{y}_S-\mu_y) \\
&= h^2\,S
\end{aligned} \qquad \textbf{Eq. 5.12}$$

Note that h^2 S is the standard formula for gain from mass selection (Falconer 1981), and that the formula was derived from a linear regression approach (i.e., Eqs. 5.1-5.4). The general equation, $\hat{g} = [Cov(g,y)/Var(y)]\ (y-\alpha)$, can be used to derive appropriate gain formulae for any selection situation.

5.3.3 Example 2: Prediction of genetic gain from mass selection

Consider the five trees described in Example 1 above. Assume we wish to select the best two for some purpose, say to be established in a two-clone orchard. Since the same heritability applies to all candidates, it is irrelevant whether selection is on the basis of phenotypic or predicted genetic values. For example, if we rank and select from the five trees on the basis of phenotype, we select the 840 and 720 cm trees. If we rank and select on the basis of predicted breeding value, we select the same trees, those with $\hat{g}$ equal to +37.68 and +18.84 (Figure 1). Thus, either approach results in the same amount of genetic gain (the offspring of a cross of these two parents would be expected to be (37.68 + 18.84)/2 = 28.26 cm taller than average). However, a different heritability may apply to some candidates, say because mass selection was being done in two plantations, one that was on an extremely uniform site, the other on a highly variable site. A large phenotypic deviation on the variable site would not indicate as much genetic superiority as an equally large deviation on the uniform site. Thus, to maximize genetic gain, it would be important to rank and select all trees on the basis of their predicted genetic value using the appropriate heritability.

5.3.4 Standardization and assumption of normality

When the observed data y have been standardized, the gain formula for mass selection can be written differently than it appears in Eq. 5.12. Any random variable y with $E(y) = \mu_y$ and $Var(y) = \sigma^2_y$ can be transformed into a random variable z with $E(z) = 0$ and $Var(z) = 1$. The transformation is

$$z = (y-\mu_y)/\sigma_y.$$

Then both 1) observed deviations ($y-\mu_y$) and 2) the selection differential (S) can be expressed as functions of standardized deviates (z):

$$1.\ (y-\mu_y) = (y-\mu_y)\,(\sigma_y/\sigma_y) = z\sigma_y$$

$$2.\ S = (\bar{y}_S-\mu_y) = (\bar{y}_S-\mu_y)\,(\sigma_y/\sigma_y) = i\sigma_y$$

where $i = (\bar{y}_S-\mu_y)/\sigma_y$, the standardized selection intensity (Falconer 1981). Thus, from Eq. 5.10, gain from mass selection is predicted as

$$\Delta\hat{g} = ih^2\sigma_y$$

which is another familiar formula (Falconer 1981, Namkoong et al. 1966, Shelbourne 1969). The observed y is often assumed to be distributed normally ($y \sim N(\mu_y, \sigma^2_y)$). Then if a breeder has some idea of the proportion of individuals to be selected, the standardized selection intensity can be estimated from tables constructed for this purpose (Fisher and Yates 1953, Falconer 1965, Becker 1975). Once the data (y) are in hand however, S is known and it is not necessary to use standardized intensity to predict gain.

5.4 Parental Selection: Both Parents Selected

The term parental selection refers here to the selection of trees based on the performance of offspring in progeny tests, either open-pollinated (OP) or cross-pollinated (CP). An example of this type of selection is roguing of some parents from a seed orchard on the basis of the performance of their offspring in progeny tests. Gain predictions are then specific to the incremental gain achieved by progeny testing and roguing, i.e., the incremental gain made from planting seed from the rogued above that from the unrogued orchard. The gain prediction would not take into account the genetic superiority already existing in the orchard, e.g., the gain achieved by establishing the orchard with ramets of genotypes from mass selection in natural stands.

For simplicity, we assume that progeny tests are established with open-pollinated seedlings in randomized complete block designs. A linear model for an individual tree observation from a progeny test is

$$y_{ijkl} = \mu_y + E_i + B_{ij} + f_k + fe_{ik} + p_{ijk} + e_{ijkl} \qquad \textbf{Eq. 5.13}$$

where,

μ_y = a fixed general mean,

E_i = fixed effect of ith test environment, $i = 1, 2, \ldots t$,

B_{ij} = fixed effect of jth block in i^{th} test, $j = 1, 2, \ldots b_i$,

f_k = random interaction effect of kth family, $k = 1, 2, \ldots s_i$, $E(f_k) = 0$, $Var(f_k) = \sigma^2_f$,

fe_{ik} = random effect of kth family in i^{th} test, $E(fe_{ik}) = 0$, $Var(fe_{ik}) = \sigma^2_{fe}$,

p_{ijk} = random plot error of kth family in j^{th} block of ith test, $E(p_{ijk}) = 0$, $Var(p_{ijk}) = \sigma^2_p$,

e_{ijkl} = random tree error of lth tree in ijk^{th} plot, $E(e_{ijkl}) = 0$, $Var(e_{ijkl}) = \sigma^2_e$,

the covariances between all pairs of factors are assumed zero, $\Sigma E_i = 0$, and ΣB_{ij} in any one test $= 0$.

The objective is to predict a breeding value for the parent of the kth open-pollinated (assumed to be half-sib) family. Breeding values (g) are random variables with E(g) = 0, and Var(g) = σ^2_a (i.e., the additive genetic variance). Each individual in the kth half-sib family receives half of its genes from the kth parent, thus the family effect associated with the kth family equals half of the breeding value of the kth parent, i.e.

$$\begin{aligned} f_k &= 1/2\ g_k \\ 2f_k &= g_k \end{aligned} \qquad \textbf{Eq. 5.14}$$

From Eq. 5.14, the family variance equals 1/4 of the additive genetic variance:

$$\begin{aligned} \mathrm{Var}(f_k) &= \mathrm{Var}(1/2\ g_k) \\ \sigma^2_f &= 1/4\ \sigma^2_a, \end{aligned} \qquad \textbf{Eq. 5.15}$$

Note that $f_k = 1/2g_k$ and $\sigma^2_f = 1/4\sigma^2_a$ are true only for true half-sib families. There are several conditions that must be met for open-pollinated families to approximate true half-sib families (Squillace 1974; Sorensen and White 1988). Throughout this chapter, we assume that open-pollinated families are half-sib families.

To predict the breeding value of the kth parent using the half-sib family mean ($\bar{y}_{..k.}$) across all tests as the observational unit, the proper form of Eq. 5.4 is

$$\hat{g}_k = [\mathrm{Cov}(\bar{y}_{..k.}, g_k)/\mathrm{Var}(\bar{y}_{..k.})]\ (\bar{y}_{..k.} - \alpha) \qquad \textbf{Eq. 5.16}$$

In terms of the linear model, the half-sib family mean is

$$\bar{y}_{\cdot\cdot k\cdot} = (\sum_i^t \sum_j^b \sum_l^n y_{ijkl})/tbn \qquad \textbf{Eq. 5.17}$$

$$= [\sum_i^t \sum_j^b \sum_l^n (\mu_y + E_i + B_{ij} + f_k + fe_{ik} + p_{ijk} + e_{ijkl})]/tbn$$

$$= \mu_y + f_k + (fe_{\cdot k}/t) + (p_{\cdot\cdot k}/tb) + (e_{\cdot\cdot k\cdot}/tbn)$$

By substituting the linear model for the observational unit $\bar{y}_{\cdot\cdot k\cdot}$ into the terms of Eq. 5.16, we can express that equation in terms of meaningful genetic parameters. For example, the covariance between the kth family mean and the breeding value of the kth parent is

$$\begin{aligned} Cov(\bar{y}_{\cdot\cdot k\cdot}, g_k) &= Cov(\mu_y + f_k + fe_{\cdot k}/t + p_{\cdot\cdot k}/tb + e_{\cdot\cdot k\cdot}/tbn,\ g_k) \\ &= Cov(f_k,\ g_k) \qquad \textbf{Eq. 5.18} \\ &= Cov(1/2 g_k,\ g_k) \\ &= 1/2\ \sigma^2_a \\ &= 2\ \sigma^2_f \end{aligned}$$

In Eq. 5.4, $\alpha = E(y)$, i.e., the expected value of the observational unit being used for selection. In this case, the expected value of the half-sib family mean is

$$\begin{aligned} E(\bar{y}_{\cdot\cdot k\cdot}) &= \mu_y + f_k + (fe_{\cdot k}/t) + (p_{\cdot\cdot k}/tb) + (e_{\cdot\cdot k\cdot}/tbn) \qquad \textbf{Eq. 5.19} \\ &= \mu_y \end{aligned}$$

since all other effects in the linear model are random variables with expected value of zero. The variance of the observed family mean $\bar{y}_{\cdot\cdot k\cdot}$ is

$$\text{Var}(\bar{y}_{\cdot\cdot k\cdot}) = \text{Var}(f_k + fe_{\cdot k}/t + p_{\cdot\cdot k}/tb + e_{\cdot\cdot k\cdot}/tbn) = \sigma^2_f + \sigma^2_{fe}/t + \sigma^2_p/tb + \sigma^2_e/tbn \qquad \textbf{Eq. 5.20}$$

Substituting the above into Eq. 5.16, the predicted breeding value for the kth parent is

$$\hat{g}_k = [2\sigma^2_f/(\sigma^2_f + \sigma^2_{fe}/t + \sigma^2_p/tb + \sigma^2_e/tbn)]\ (\bar{y}_{\cdot\cdot k\cdot} - \alpha) = 2\ h^2_f\ (\bar{y}_{\cdot\cdot k\cdot} - \alpha) \qquad \textbf{Eq. 5.21}$$

where h^2_f, the family heritability is defined as $\sigma^2_f/(\sigma^2_f + \sigma^2_{fe}/t + \sigma^2_p/tb + \sigma^2_e/tbn)$. If N parents are selected, the predicted gain is the average of the predicted breeding values:

$$\Delta\hat{g} = \Sigma\ \hat{g}_k/N = 2\ h^2_f\ S_f \qquad \textbf{Eq. 5.22}$$

where S_f = mean of $(\bar{y}_{\cdot\cdot k\cdot} - \alpha)$ across the selected families = selection differential on a family mean basis. Again, this common formula (Namkoong et al. 1966, Shelbourne 1969) was derived from the general regression approach in Eq. 5.4.

5.4.1 Example 3: Parental selection

Assume that we wish to predict the breeding values for 10-year height of coastal Douglas-fir parents using height data from 10-year-old OP progeny tests. For each parent, its half-sib family is planted in t = 6 = number of tests, and each test has b = 4 = number of blocks, with n = 3 = number of trees per block. Assume the genetic parameters given in Table 5.1 for Region B apply to all tests. Height has an additive genetic variance = σ^2_a = 2216 at 10 years. Also, the observational unit is the family mean across all tests $(\bar{y}_{\cdot\cdot k\cdot})$, and Eqs.

Table 5.1 Assumed genetic parameters for 10-year-old coastal Douglas-fir height (in cm) for three hypothetical breeding regions.[1]

Region	μ	σ^2_f	σ^2_{fe}	σ^2_p	σ^2_e	σ^2_a	σ^2_T	h^2
A	600	554	392	245	7673	2216	8864	0.250
B	600	554	392	396	12796	2216	14138	0.157
C	600	554	392	396	12796	2216	14138	0.157

[1]Parameters for Regions B and C were estimated from actual data: 12 sets of 30 open-pollinated families/set across 8 randomized complete block progeny tests. Parameters for Region A are hypothetical modifications of those from Regions B and C. Genetic correlations between breeding regions are hypothetical and developed to illustrate computational principles; they are not based on actual data.

$r_{g:A,B} = 0.6$, $Cov(g^A_k, g^B_k) = 1329.6$ $\quad$ $r_{g:A,A} = 1.0$, $Cov(g^A_k, g^A_k) = 2216$
$r_{g:A,C} = 0.4$, $Cov(g^A_k, g^C_k) = 886.4$ $\quad$ $r_{g:B,B} = 1.0$, $Cov(g^B_k, g^B_k) = 2216$
$r_{g:B,C} = 0.7$, $Cov(g^B_k, g^C_k) = 1551.2$ $\quad$ $r_{g:C,C} = 1.0$, $Cov(g^C_k, g^C_k) = 2216$

5.19 and 5.20 are used to calculate $E(\bar{y}_{\cdot\cdot k\cdot}) = \alpha = 600$ and $Var(\bar{y}_{\cdot\cdot k\cdot}) = 813.56$. Eq. 5.18 is used to calculate $Cov(\bar{y}_{\cdot\cdot k\cdot}, g_k) = 1/2\ \sigma^2_a = 1108$.

Thus, for each parent, the breeding value is predicted using Eq. 5.16:

$$\begin{aligned} \hat{g}_k &= [Cov(\bar{y}_{\cdot\cdot k\cdot}, g_k)/Var(\bar{y}_{\cdot\cdot k\cdot})]\ (\bar{y}_{\cdot\cdot k\cdot} - \alpha) \\ &= (1108/813.56)\ (\bar{y}_{\cdot\cdot k\cdot} - \alpha) \\ &= 1.3619\ (\bar{y}_{\cdot\cdot k\cdot} - \alpha). \end{aligned}$$

The above formula is also equivalent to Eq. 5.21 where the coefficient on the phenotypic deviation equals twice the family heritability ($2\ h^2_f$)

since $h^2_f = 0.681$. Now assume there are three half-sib families (A, B, and C) which are 1, 1.5 and 2 phenotypic standard deviations above the mean (standard deviation on a family mean basis is $813.56^{1/2} = 28.52$). For those families, $\bar{y}_{..k.}$, $(\bar{y}_{..k.} - \alpha)$, and g_k are presented in Table 5.2.

Table 5.2 Family mean across six tests ($\bar{y}_{..k.}$), family mean deviation from expected value ($\bar{y}_{..k.} - \alpha$), and predicted parental breeding value ($\hat{g}_k$) for three hypothetical Douglas-fir parents. Genetic parameters for Region B (Table 5.1) are assumed.

Phenotypic or predicted genetic value	Family		
	A	B	C
$\bar{y}_{..k.}$	628.52	642.78	657.04
$(\bar{y}_{..k.} - \alpha)$	28.52	42.78	57.04
$\hat{g}_k$	38.84	58.26	77.68

Note that the predicted breeding value for all parents is greater than the observed phenotypic deviation. Although it may appear that the predicted breeding values are not being regressed toward the mean of zero, the predictions in fact *are* regressed. It may help to gain a heuristic understanding of this by considering Eq. 5.21, $\hat{g}_k = 2\, h^2_f\, (\bar{y}_{..k.} - \alpha)$. The observed phenotypic deviation, $\bar{y}_{..k.} - \alpha$, is regressed toward zero by multiplying by the family heritability, since some of that deviation is likely to be due to environmental and non-additive genetic effects. For example, for family B with $\bar{y}_{..k.} - \alpha = 42.78$, the predicted family effect ($\hat{f}_k$) is 0.681 (42.78) = 29.13, thus $\hat{f}_k$ was predicted by regressing the phenotypic deviation toward zero. However, to predict the breeding value of the kth parent (which contributed a random sample of half its genes to each offspring in the half-sib family), we multiply $\hat{f}_k$ by 2 to calculate $\hat{g}_k$. The correlation between the predicted and true breeding values is calculated using Eqs. 5.6 and 5.8: $\mathrm{Var}(\hat{g}) = 1108^2/813.56 = 1509$, and given that $\mathrm{Var}(g) = 2216$, $\mathrm{Corr}(g,\hat{g}) = 0.83$.

5.5 Parental Selection: One Parent Selected

In this case of parental selection, only one of the two parents is selected as superior, so the selection is essentially for particular half-sib families (and for this reason is also called family selection). An example of this type of selection is *directed seed collection* in which progeny test data are used to identify superior parents in an unrogued seed orchard. If seed for operational reforestation is collected only from these superior parents (but the orchard is not rogued so that all parents contribute to the male gamete pool), then gains will be higher than from seed collected from all parents in the unrogued orchard, but not as high as from the rogued orchard (in which both the male and female gamete pools are upgraded from the progeny test data).

In this situation we wish to use a half-sib family mean ($\bar{y}_{..k.}$) to predict an unknown genetic value, one-half of a parental breeding value (1/2 g_k), which is the half-sib family effect (f_k). This is the genetic gain which will be realized from future plantings. Remember that this is just the incremental gain (above that of the initial orchard) due to identifying the better female parents from progeny test data. In this case,

$$\begin{aligned}
Cov(\bar{y}_{..k.}, f_k) &= Cov(f_k + fe_{.k}/t + p_{..k}/tb + e_{..k.}/tbn,\ f_k) \\
&= Cov(f_k, f_k) \\
&= \sigma^2_f \\
&= 1/4\ \sigma^2_a
\end{aligned}$$

To predict 1/2 g_k, this covariance is only one-half as large as the covariance when the trait being predicted was g_k (Eq. 5.18). Thus, following the derivation for the two parent case and assuming N parents selected, the genetic gain formula for family selection reduces to

$$\Delta\hat{g} = \Sigma\ \hat{g}_k/N = h^2_f\ S_f$$

Since selection is only for one parent (the female in this case) genetic gain is one-half as large as when selection is for both parents (Eq. 5.22).

5.6 Indirect Selection

To this point only direct selection has been discussed, e.g., predicting breeding values for 10-year height from measurements of 10-year height. Indirect selection is the selection of individuals for one trait on the basis of measurements on a second trait. An example in forestry would be any type of early selection based on growth at 1 year, 5 years, 10 years, etc., for the trait of volume per acre at harvest. Indirect selection is used whenever measurement of the trait of interest is expensive, impractical, or impossible.

We use the general prediction formula (Eq. 5.4) to predict a genetic value for one trait from measurements on a different trait. A linear model for two traits $y_{i:1}$ and $y_{i:2}$ is:

$$y_{i:1} = \mu_{y:1} + g_{i:1} + e_{i:1} \qquad \textbf{Eq. 5.23}$$

where,

$y_{i:1}$ = phenotypic observation on i^{th} tree, $E(y_{i:1})=\mu_{y:1}$, $Var(y_{i:1}) = \sigma^2_{y1}$,

$g_{i:1}$ = breeding value for the ith tree, $E(g_{i:1})=0$, $Var(g_{i:1})=\sigma^2_{a1}$,

$e_{i:1}$ = error for the ith tree, $E(e_{i:1})=0$, $Var(e_{i:1})=\sigma^2_{e1}$, and there is no covariance between terms, $\sigma^2_{y1} = \sigma^2_{a1} + \sigma^2_{e1}$.

For the second trait,

$$y_{i:2} = \mu_{y:2} + g_{i:2} + e_{i:2} \qquad \textbf{Eq. 5.24}$$

where,

$y_{i:2}$ = phenotypic observation on i^{th} tree, $E(y_{i:2})=\mu_{y:2}$, $Var(y_{i:2})=\sigma^2_{y2}$,

$g_{i:2}$ = breeding value for the ith tree, $E(g_{i:2})=0$, $Var(g_{i:2})=\sigma^2_{a2}$,

$e_{i:2}$ = error for the ith tree, $E(e_{i:2})=0$, $Var(e_{i:2})=\sigma^2_{e2}$, and there is no covariance between terms, $\sigma^2_{y2} = \sigma^2_{a2} + \sigma^2_{e2}$.

The genetic relationship between traits 1 and 2 allows use of measurements on one trait to predict the genetic value for the other trait. The additive genetic covariance between the two traits is $Cov(g_{i:1},g_{i:2}) = \sigma_{g1,g2} = \sigma_{a1,a2}$. To predict the breeding value for trait 1 ($g_{i:1}$) as a function of measurements on trait 2 ($y_{i:2}$), use Eq. 5.4 (all i subscripts have been deleted for simplicity):

$$\hat{g}_1 = [Cov(g_1,y_2)/Var(y_2)]\ (y_2 - \alpha_2) \qquad \textbf{Eq. 5.25}$$

$$\hat{g}_1 = \frac{\mathrm{Cov}(g_1,y_2)[\mathrm{Var}(g_1)]^{1/2}[\mathrm{Var}(g_2)]^{1/2}[\mathrm{Var}(y_1)]^{1/2}}{[\mathrm{Var}(y_2)][\mathrm{Var}(g_1)]^{1/2}[\mathrm{Var}(g_2)]^{1/2}[\mathrm{Var}(y_1)]^{1/2}}(y_2-\alpha_2)$$

$$= \frac{\mathrm{Cov}(g_1,g_2)\ [\mathrm{Var}(g_1)]^{1/2}[\mathrm{Var}(g_2)]^{1/2}[\mathrm{Var}(y_1)]^{1/2}}{[\mathrm{Var}(g_1)\ \mathrm{Var}(g_2)]^{1/2}\ [\mathrm{Var}(y_1)]^{1/2}[\mathrm{Var}(y_2)]^{1/2}[\mathrm{Var}(y_2)]^{1/2}}(y_2-\alpha_2)$$

$$= \mathrm{Corr}(g_1,g_2)\ h_1 h_2 (\sigma_{y1}/\sigma_{y2})\ (y_2-\alpha_2)$$

If N individuals are selected, the genetic gain from indirect selection is

$$\Delta\hat{g}_1 = [\mathrm{Cov}(g_1,y_2)/\mathrm{Var}(y_2)]\ S_2 \qquad \textbf{Eq. 5.26}$$
$$= \mathrm{Corr}(g_1,g_2)\ h_1\ h_2\ \sigma_{y1}\ i$$

where,

$S_2 = (\bar{y}_{S2}-\alpha_2)$, the selection differential for trait 2.

Note Eq. 5.26 is an often used gain formula for indirect selection (Falconer 1981), and it is the more general form of the gain formula from mass selection. If $y_1 = y_2$, then $\mathrm{Corr}(g_1,g_2) = 1$, $h_1 \bullet h_1 = h^2_1$, therefore $\Delta\hat{g}_1 = i\ h^2_1\ \sigma_{y1}$.

It is possible to think of all selection as indirect selection. Assume that we are interested in improving the genetic value for trait 1, but that we will actually rank candidates and make selection based on some other trait 2. If we plan to select some constant number or proportion of individuals, then we can consider i, h_1, σ_{y1} in Eq. 5.26 as constants. Then the only factors that affect the genetic gain are the heritability of the observed trait y_2 and the genetic correlation between traits 1 and 2. Specifically $\Delta\hat{g}$ will be proportional to the product of the heritability and the genetic correlation, i.e.,

$$\Delta\hat{g}_1 \text{ is proportional to } (h_2 \text{ Corr}(g_1,g_2)).$$

Thus, gain in a particular trait from indirect selection (for a given selection intensity) depends on two factors:

1) Is the observed information reliable, i.e., is it due to genetic effects? This is measured by h_2.
2) Is the observed information related to the trait of interest? This is measured by $\text{Corr}(g_1,g_2)$.

Whenever genetic worth or genetic gain is predicted using the general linear regression approach (Eqs. 5.3 or 5.4), these factors are accounted for by specifying the appropriate variances and covariances. This becomes particularly appealing in situations where multiple observations (say on several different traits and/or kinds of relatives) are used to predict a genetic worth. In the following sections we present an introduction to prediction of genetic values using multiple traits as observations, and consider some specific types of multiple trait situations as examples.

5.7 Multiple Traits

Often in tree improvement, multiple sources of information are available to the breeder who wishes to evaluate a particular individual.

For example, for a particular tree k, there may be phenotypic measurements on that individual, measurements on its siblings and/or its parents and/or its offspring in OP and CP progeny tests, various types of ancestral data, etc. To increase the chances of maximizing genetic gain, the breeder should use all data available. In another example, we could be interested in some type of genetic worth aggregated across a number of different traits with varying economic importance, e.g., volume growth, disease resistance, stem straightness, branch diameter, wood quality. Thus, the breeder would need to combine measurements on all these traits into a single prediction of aggregate genetic worth. Lastly, multiple progeny test measurements of the 'same' trait could be treated as different traits. This would be useful if the measurements from different tests had different genetic parameters, which could be the case if tests were of different precision, on different types of sites, or in different breeding zones.

The concepts discussed throughout the remainder of this chapter are similar to those presented in preceding sections. However, since multiple observations are being used to predict genetic values, it becomes convenient to utilize matrix algebra notation, which is a useful way to summarize lengthy mathematical formulae. In this chapter, vectors will be denoted with **boldface** lowercase letters, while matrices will be denoted with boldface uppercase letters. Although it should be possible to understand many of the essential concepts in the following sections, some readers may wish to take advantage of one of several excellent introductions to matrix notation including Neter and Wasserman 1974 (Ch. 6), Searle 1966, and Noble and Daniel 1977.

5.7.1 General prediction formula for multiple traits

The problem is to predict a genetic value for the kth tree ($\hat{g}_k$) as some function of all relevant observed data. A linear equation for using n measurements (which could be measurements on different traits, relatives, ages, environments, etc.) to predict a single genetic value $\hat{g}_k$ is

$$\hat{g}_k = a + b_1y_1 + b_2y_2 + ... + b_ny_n.$$

In matrix notation, this can be written as $\hat{g}_k = a + \beta'y$ where a is an intercept constant, β is a vector of coefficients b_i, and y is a vector of observations, i.e a series of random variables y_i. If the first and second moments associated with the random variables g, y_1, y_2, ... , y_n are assumed known, then a formula can be derived for calculating β such that $\hat{g}$ is a good predictor of the true genetic value. Define Q = $E(g-\hat{g})^2$, and solve for β to minimize Q. This is essentially the same process used in the derivation earlier in this chapter, with the only difference being that then a single observation was used to predict the genetic value, whereas now multiple observations are used. After considerable matrix algebra (see White and Hodge 1989, chapter 4), it can be shown that

$$\hat{g} = \tau + c'V^{-1}(y-\alpha) \qquad \textbf{Eq. 5.29}$$

where,

c = Cov(g,y),
V = Var(y), and
α = E(y).

Since the observed y contains multiple observations (e.g., individual tree measurements on a number of traits, or family means from different kinds of tests), α is a vector of expected values of all the different observations in y, V is a matrix of variances and covariances of all observations, and **c** is a vector of covariances between all observations and the genetic value being predicted. (Actually, it is possible to predict more than one genetic value, or an aggregate genetic value in a single matrix equation. In that case **C** is a matrix of covariances of all observations with all genetic values of interest. In this chapter, we restrict ourselves to prediction of single genetic values, thus **c** is a vector.)

When there is only a single observation (e.g., an individual tree measurement or a family mean as in previous sections), **c**, **V**, and $\boldsymbol{\alpha}$ all become scalars, and Eq. 5.29 reduces to Eq. 5.3, or Eq. 5.4 assuming $\tau = 0$. Thus, the essential components of the Eqs. 5.3 and 5.29 are identical (note that multiplying by $\mathbf{V}^{-1}$ is the matrix equivalent of dividing by Var(y)), and Eq. 5.29 has the same attractive statistical properties discussed earlier for Eq. 5.3. Similarly, Eqs. 5.6, 5.7, and 5.8 have their matrix equivalents in Eqs. 5.30, 5.31 and 5.32, respectively.

$$\mathrm{Var}(\hat{g}) = \mathbf{c}'\mathbf{V}^{-1}\mathbf{c} \quad \textbf{Eq. 5.30}$$

$$\mathrm{Cov}(g,\hat{g}) = \mathbf{c}'\mathbf{V}^{-1}\mathbf{c} = \mathrm{Var}(\hat{g}) \quad \textbf{Eq. 5.31}$$

$$\mathrm{Corr}(g,\hat{g}) = [\mathrm{Var}(\hat{g})/\mathrm{Var}(g)]^{1/2} \quad \textbf{Eq. 5.32}$$

Comparing Eqs. 5.6 and 5.30, **c'c** is the matrix equivalent of $[\mathrm{Cov}(g,y)]^2$. The correlation between the predicted and true genetic values ($\mathrm{Corr}(g,\hat{g})$) has the same meaning and is interpreted in the same manner as described earlier. Following are some applications of these formulae to multiple trait situations, illustrated with specific numerical examples.

5.8 Family and Within Family Selection

In many breeding programs, it is common to do some kind of family and within family selection. This involves the selection of the best families (on the basis of progeny test information) and the best individuals within families (on the basis of individual tree phenotype). We are interested in the genetic value of an individual tree, and will use two observations, a family mean and individual tree measurement, to predict that genetic value.

5.8.1 Forming the y and E(y) vectors, and V matrix

Assume there are open-pollinated, randomized complete block genetic tests, and that the linear model in Eq. 5.13 applies. The data vector y contains two observations: the kth family mean across all tests and the individual tree observation. Vectors y and E(y) are

$$\mathbf{y} = \begin{bmatrix} \text{ind. obs.} \\ \text{fam. mean} \end{bmatrix} = \begin{bmatrix} y_{ijkl} \\ \bar{y}_{\cdot\cdot k\cdot} \end{bmatrix}$$

$$E(\mathbf{y}) = \begin{bmatrix} E(y_{ijkl}) \\ E(\bar{y}_{\cdot\cdot k\cdot}) \end{bmatrix}$$

$$= \begin{bmatrix} \mu_y + E_i + B_{ij} \\ \mu_y \end{bmatrix}$$

In this case, with two observations in y, the 2 x 2 matrix **V** containing the variances and covariances of those observations (i.e., the second moments) is:

$$\mathbf{V} = \begin{bmatrix} Var(y_{ijkl}) & Cov(y_{ijkl}, \bar{y}_{\cdot\cdot k\cdot}) \\ Cov(y_{ijkl}, \bar{y}_{\cdot\cdot k\cdot}) & Var(\bar{y}_{\cdot\cdot k\cdot}) \end{bmatrix}$$

The diagonal elements of this symmetric matrix are the variances of the two observations, while both off-diagonal elements are the covariance between the two. The variance of family means ($Var(\bar{y}_{\cdot\cdot k\cdot})$) has already been derived (Eq. 5.20). By substituting the linear model (Eq. 5.13) for the individual tree observation, and Eq. 5.17 for the family mean, expressions for the remaining two types of second moments are:

$$\begin{aligned} Var(y_{ijkl}) &= Var(\mu + E_i + B_{ij} + f_k + fe_{ik} + p_{ijk} + e_{ijkl}) \\ &= \sigma^2_f + \sigma^2_{fe} + \sigma^2_p + \sigma^2_e \end{aligned} \qquad \textbf{Eq. 5.33}$$

$$\begin{aligned} Cov(y_{ijkl}, \bar{y}_{\cdot\cdot k\cdot}) &= Cov(\mu + E_i + B_{ij} + f_k + fe_{ik} + p_{ijk} + e_{ijkl}, \\ &\quad \mu + f_k + fe_{\cdot k}/t + p_{\cdot\cdot k}/tb + e_{\cdot\cdot k\cdot}/tbn) \end{aligned} \qquad \textbf{Eq. 5.34}$$

$$\begin{aligned} &= Cov(f_k, f_k) + Cov(fe_{ik}, fe_{ik})/t + Cov(p_{ijk}, p_{ijk})/tb \\ &\quad + Cov(e_{ijkl}, e_{ijkl})/tbn \end{aligned}$$

$$= \sigma^2_f + \sigma^2_{fe}/t + \sigma^2_p/tb + \sigma^2_e/tbn$$

$$= Var(\bar{y}_{\cdot\cdot k\cdot})$$

5.8.2 Forming the c vector

The vector **c** contains covariances between the measured observations and the breeding value of the ijkl[th] tree (g_{ijkl}). Since there are only two observations in **y**, there will be two elements in **c**.

$$\mathbf{c} = \begin{bmatrix} Cov(y_{ijkl}, g_{ijkl}) \\ Cov(\bar{y}_{..k.}, g_{ijkl}) \end{bmatrix}$$

To determine what these covariances are, genetic theory is used to define the breeding value g_{ijkl} as a function of terms in the linear model for observations. Let g_{ijkl} be

$$g_{ijkl} = f_k + w_{ijkl} \qquad \textbf{Eq. 5.35}$$

where,

g_{ijkl} = the breeding value of the lth individual in the kth family in the jth block of the ith test,

f_k = random effect of the kth family, $E(f_k)=0$, $Var(f_k) = \sigma^2_f = 1/4\ \sigma^2_a$,

w_{ijkl} = random additive genetic effect of the lth individual in the jth block of the kth family in the ith test, $E(w_{ijkl})=0$, $Var(w_{ijkl}) = 3\sigma^2_f = 3/4\ \sigma^2_a$.

The term w_{ijkl} accounts for the part of the breeding value of the ijkl[th] individual which cannot be accounted for by f_k, the average effect of the kth parent. In the original linear model (Eq. 5.13), the term e_{ijkl} contains w_{ijkl}, the within-family genetic deviation, as well as other random error effects. For simplicity, we will ignore the fact that p_{ijk} contains a small portion of the within-family genetic deviation (see Namkoong 1979, p77). We can expand the linear model in Eq. 5.13 by defining e_{ijkl} as follows:

$$e_{ijkl} = w_{ijkl} + e^{*}_{ijkl} \qquad \textbf{Eq. 5.36}$$

where e^{*}_{ijkl} contains both random non-additive genetic effects and microsite environmental effects associated with l^{th} individual in the ijk^{th} plot.

The elements of **c** can now be expressed in terms of meaningful genetic parameters:

$$\begin{aligned}
\text{Cov}(y_{ijkl}, g_{ijkl}) &= \text{Cov}(\mu + E_i + B_{ij} + f_k + fe_{ik} + p_{ijk} + e_{ijkl}, g_{ijkl}) \\
&= \text{Cov}(f_k, g_{ijkl}) + \text{Cov}(e_{ijkl}, g_{ijkl}) \\
&= \text{Cov}(f_k, f_k+w_{ijkl}) + \text{Cov}(e^{*}_{ijkl} + w_{ijkl}, f_k+w_{ijkl}) \\
&= \text{Cov}(f_k, f_k) + \text{Cov}(w_{ijkl}, w_{ijkl}) \qquad \textbf{Eq. 5.37} \\
&= \sigma^2_f + 3\sigma^2_f \\
&= \sigma^2_a.
\end{aligned}$$

$$\begin{aligned}
\text{Cov}(\bar{y}_{..k.}, g_{ijkl}) &= \text{Cov}(\mu + f_k + fe_{.k}/t + p_{..k}/tb + e_{..k.}/tbn, g_{ijkl}) \\
&= \text{Cov}(f_k, g_{ijkl}) + \text{Cov}(e_{..k.}/tbn, g_{ijkl}) \\
&= \text{Cov}(f_k, f_k+w_{ijkl}) + \text{Cov}(e^{*}_{..k.}/tbn + w_{..k.}/tbn, f_k+w_{ijkl}) \\
&= \text{Cov}(f_k, f_k) + (1/tbn)\,\text{Cov}(w_{ijkl}, w_{ijkl}) \qquad \textbf{Eq. 5.38} \\
&= \sigma^2_f + (3/tbn)\,\sigma^2_f \\
&= 1/4\ \sigma^2_a + (3/4\ tbn)\,\sigma^2_a.
\end{aligned}$$

5.8.3 Example 4: Prediction of breeding values from family and within-family selection

Consider again the 10-year-old coastal Douglas-fir genetic tests described in Example 3. Half-sib families are planted in 6 tests, with 3 trees/block and 4 blocks in each test. The genetic parameters given for Region B, Table 5.1, apply to these tests. Given these parameters, the following second moments form **V** and **c**:

$Var(y_{ijkl}) = 14138$	(from Eq. 5.33)
$Var(\bar{y}_{\cdot\cdot k\cdot}) = 813.56$	(from Eq. 5.20)
$Cov(y_{ijkl}, \bar{y}_{\cdot\cdot k\cdot}) = 813.56$	(from Eq. 5.34)
$Cov(y_{ijkl}, g_{ijkl}) = 2216$	(from Eq. 5.37)
$Cov(\bar{y}_{\cdot\cdot k\cdot}, g_{ijkl}) = 577.08$	(from Eq. 5.38)

$$\mathbf{V} = \begin{bmatrix} 14138 & 813.56 \\ 813.56 & 813.56 \end{bmatrix} \qquad \mathbf{c} = \begin{bmatrix} 2216 \\ 577.08 \end{bmatrix}$$

Assuming that $\tau = 0$, Eq. 5.29 reduces to $\hat{g}_{ijkl} = \mathbf{c'V^{-1}(y\text{-}\alpha)}$. The predicted breeding values in this case can then be thought of as deviations from the expected breeding value τ. By defining a vector $\boldsymbol{\beta}'$ as $\mathbf{c'V^{-1}}$, then for these parameters, $\boldsymbol{\beta}' = [\ 0.1230 \quad 0.5863]$. The vector $\boldsymbol{\beta}'$ contains the coefficients that are multiplied by the deviations of the observed data from their expected values. Thus,

$$\hat{g}_{ijkl} = 0.1230\ [(y_{ijkl} - E(y_{ijkl})] + 0.5863\ [(\bar{y}_{\cdot\cdot k\cdot} - E(\bar{y}_{\cdot\cdot k\cdot})]$$

This is a selection index equation which is used repeatedly to predict genetic values for every individual tree. Note that for the Douglas-fir parameters assumed in this case there is considerably more weight placed on the family mean than the individual tree phenotype.

We can also use **V** and **c** to estimate the correlation between true and predicted breeding values for the individual trees; $Corr(g,\hat{g}) = 0.53$ from Eq. 5.32. Note that both the index coefficients in $\boldsymbol{\beta}'$ and $Corr(g,\hat{g})$ were calculated without any actual data (**y**); we only used the parameters that were assumed to apply to data from this specified type of inference population (i.e., the assumed parameters specify the population structure).

Recall that the standard deviation of family means is $(813.56)^{1/2}$ = 28.52. Consider the same three half-sib families (A, B, and C) discussed in Example 3 (Section 5.4.1), which were 1, 1.5 and 2 standard deviations (on a family mean basis) above the mean. Assume that the best individual within each of these families is 230.86 cm above the family mean (the phenotypic standard deviation within a half-sib family is 115.43; 230.86 is two standard deviations above the family mean). For those three individuals, the deviation of the individual tree value above the expected value for their block means (i.e., $y_{ijkl} - \alpha_{ijkl}$), and the deviation of the half-sib family above the overall mean (i.e., $\bar{y}_{..k.} - \alpha_{..k.}$) are used in Eq. 5.29 to predict $\hat{g}_{ijkl}$, the breeding value for the ijkl[th] tree (Table 5.3).

Table 5.3 Comparison of family and within-family selection (forward selection) and parental selection (backward selection) for three hypothetical Douglas-fir families.[1]

Phenotypic or predicted genetic value	Family A	Family B	Family C
$(y_{ijkl} - \alpha_{ijkl})$	259.38	273.64	287.90
$(\bar{y}_{..k.} - \alpha_{..k.})$	28.52	42.78	57.04
$\hat{g}_{ijkl}$	48.63	58.74	68.85
$\hat{g}_k$	38.84	58.26	77.68

[1]Values presented are individual tree deviations from expected value $(y_{ijkl} - \alpha_{ijkl})$, family mean deviation from expected value $(\bar{y}_{..k.} - \alpha_{..k.})$, predicted individual tree breeding value $(\hat{g}_{ijkl})$, and predicted parental breeding value $(\hat{g}_k)$. Genetic parameters for Region B (Table 5.1) are assumed.

5.8.4 Forward vs backward selection

The term forward selection is commonly used when the best individuals within the best families are selected to be included in a production or breeding population. A related term, backward selection, is used to describe selection of progeny tested parents to be included in a production or breeding population. When a breeder is making choices among candidates, the opportunity exists to make either forward or backward selections, or both. Consider the predicted breeding values for the best forward selections in the half-sib families A, B and C, compared to the predicted breeding values for the backward selections, the parents A, B, and C (Table 5.3). The best forward selection in family A has a higher $\hat{g}$ than parent A, but the best individual in family C has a lower $\hat{g}$ than parent C. With the parameters used in Examples 2 and 3, the breeding value of a parent is higher than the breeding value of the best individual in the family for approximately the top 6% of the half-sib families. This assumes that the best individual in the family is 2 standard deviations (on a within family basis) above the family mean, and in practice, this would vary from family to family. However, the general trend holds: more gain is made from backward selection of the best parents than from forward selection among their offspring. The advantage of basing selections on predicted genetic worth is that all genotypes available for consideration (both forward and backward selections) can be compared on an equal basis since genetic values for the same trait are predicted in every case.

5.8.5 Alternate approach to family and within-family selection

In the previous development of family and within-family selection, the observations in **y** were the phenotypic value (y_{ijkl}) and the overall family mean ($\bar{y}_{\cdot\cdot k\cdot}$). An alternative is to use the within-family deviation (y_{ijkl} - $\bar{y}_{\cdot\cdot k\cdot}$) in the data vector **y**. In this case, **y**, E(**y**), **V** and **c** would be

$$\mathbf{y} = \begin{bmatrix} \text{ind. dev.} \\ \text{fam. mean} \end{bmatrix} = \begin{bmatrix} y_{ijkl}, \bar{y}_{\cdot\cdot k\cdot} \\ \bar{y}_{\cdot\cdot k\cdot} \end{bmatrix}$$

$$\mathrm{E}(\mathbf{y}) = \begin{bmatrix} E_i + B_{ij} \\ \mu_y \end{bmatrix}$$

$$\mathbf{V} = \begin{bmatrix} \mathrm{Var}(y_{ijkl}\bar{y}_{\cdot\cdot k\cdot}) & \mathrm{Cov}(y_{ijkl}\bar{y}_{\cdot\cdot k\cdot}, \bar{y}_{\cdot\cdot k\cdot}) \\ \mathrm{Cov}(y_{ijkl}\bar{y}_{\cdot\cdot k\cdot}, \bar{y}_{\cdot\cdot k\cdot}) & \mathrm{Var}(\bar{y}_{\cdot\cdot k\cdot}) \end{bmatrix}$$

$$\mathbf{c} = \begin{bmatrix} \mathrm{Cov}(y_{ijkl}, \bar{y}_{\cdot\cdot k\cdot}, g_{ijkl}) \\ \mathrm{Cov}(\bar{y}_{\cdot\cdot k\cdot}, g_{ijkl}) \end{bmatrix}$$

By substituting the linear model for the various terms in **V** and **c**, the following expressions result

$$\begin{aligned} \mathrm{Var}(y_{ijkl}-\bar{y}_{\cdot\cdot k\cdot}) &= \mathrm{Var}(y_{ijkl}) + \mathrm{Var}(\bar{y}_{\cdot\cdot k\cdot}) - 2\mathrm{Cov}(y_{ijkl}, \bar{y}_{\cdot\cdot k\cdot}) \\ &= \mathrm{Var}(y_{ijkl}) - \mathrm{Var}(\bar{y}_{\cdot\cdot k\cdot}) \qquad \textbf{Eq. 5.39} \end{aligned}$$

$$\begin{aligned} Cov(y_{ijkl}-\bar{y}_{\cdot\cdot k\cdot}, \bar{y}_{\cdot\cdot k\cdot}) &= Cov(y_{ijkl}, \bar{y}_{\cdot\cdot k\cdot}) - Cov(\bar{y}_{\cdot\cdot k\cdot}, \bar{y}_{\cdot\cdot k\cdot}) \\ &= Var(\bar{y}_{\cdot\cdot k\cdot}) - Var(\bar{y}_{\cdot\cdot k\cdot}) \\ &= 0 \end{aligned} \qquad \textbf{Eq. 5.40}$$

$$Cov(\bar{y}_{\cdot\cdot k\cdot}, g_{ijkl}) = 1/4\ \sigma^2_a + (3/4\ tbn)\ \sigma^2_a. \qquad \textbf{Eq. 5.41}$$

$$\begin{aligned} Cov(y_{ijkl}-\bar{y}_{\cdot\cdot k\cdot}, g_{ijkl}) &= Cov(y_{ijkl}, g_{ijkl}) - Cov(\bar{y}_{\cdot\cdot k\cdot}, g_{ijkl}) \\ &= \sigma^2_a - [1/4\ \sigma^2_a + (3/4\ tbn)\ \sigma^2_a] \end{aligned} \qquad \textbf{Eq. 5.42}$$

Given the genetic parameters for Region B, Table 5.1, the following estimates of second moments form **V** and **c**:

$Var(y_{ijkl} - \bar{y}_{\cdot\cdot k\cdot})$	=	13324.44	(from Eq. 5.39)
$Var(\bar{y}_{\cdot\cdot k\cdot})$	=	813.56	(from Eq. 5.20)
$Cov(y_{ijkl} - \bar{y}_{\cdot\cdot k\cdot}, \bar{y}_{\cdot\cdot k\cdot})$	=	0	(from Eq. 5.40)
$Cov(y_{ijkl} - \bar{y}_{\cdot\cdot k\cdot}, g_{ijkl})$	=	1638.92	(from Eq. 5.42)
$Cov(\bar{y}_{\cdot\cdot k\cdot}, g_{ijkl})$	=	577.08	(from Eq. 5.41)

Then,

$$\mathbf{V} = \begin{bmatrix} 13324.44 & 0 \\ 0 & 813.56 \end{bmatrix} \qquad \mathbf{c} = \begin{bmatrix} 1638.92 \\ 577.08 \end{bmatrix}$$

In this case, the vector of coefficients $\boldsymbol{\beta}' = \mathbf{c}'\mathbf{V}^{-1} = [\ 0.1230 \quad 0.7093]$.

Thus,

$$\begin{aligned}\hat{g}_{ijkl} &= 0.1230\,[(y_{ijkl} - \bar{y}_{\cdot\cdot k\cdot}) - E(y_{ijkl} - \bar{y}_{\cdot\cdot k\cdot})] + 0.7093\,[(\bar{y}_{\cdot\cdot k\cdot} - E(\bar{y}_{\cdot\cdot k\cdot})] \\ &= 0.1230\,[(y_{ijkl} - E(y_{ijkl})] - 0.1230\,[(\bar{y}_{\cdot\cdot k\cdot}) - E(\bar{y}_{\cdot\cdot k\cdot})] + 0.7093\,[(\bar{y}_{\cdot\cdot k\cdot} - E(\bar{y}_{\cdot\cdot k\cdot})] \\ &= 0.1230\,[(y_{ijkl} - E(y_{ijkl})] + 0.5863\,[(\bar{y}_{\cdot\cdot k\cdot} - E(\bar{y}_{\cdot\cdot k\cdot})]\end{aligned}$$

which is identical to the index equation developed with the previous approach presented in Example 4.

Often family and within-family selection indices are approximated by using the family heritability and within-family heritability (h^2_w) as coefficients on the family mean and within-family deviations, respectively. Note that in the above equation, the coefficient 0.7093 on the family mean is approximately equal to the family heritability:

$$\begin{aligned}0.7093 &= 577.08 / 813.56 \\ &= Cov(\bar{y}_{\cdot\cdot k\cdot}, g_{ijkl}) / Var(\bar{y}_{\cdot\cdot k\cdot}) \\ &= [1/4\ \sigma^2_a + (3/4\ tbn)\ \sigma^2_a] / [\sigma^2_f + \sigma^2_{fe}/t + \sigma^2_p/tb + \sigma^2_e/tbn] \\ &\approx h^2_f\ .\end{aligned}$$

The coefficient will approach the family heritability as numbers of tests, blocks, and trees per block (t, b, and n) increase. Within-family heritability (Namkoong et al. 1966, Shelbourne 1969) is commonly calculated as:

$$h^2_w = (3/4)\ \sigma^2_a / (\sigma^2_{fe}/t + \sigma^2_p/tb + \sigma^2_e/tbn).$$

The coefficient on the within-family deviation (0.1230) equals the additive genetic variance within a half-sib family ($3/4\ \sigma^2_a$) divided by the variance of within-family deviations ($Var(y_{ijkl} - \bar{y}_{\cdot\cdot k\cdot})$), and is approximately equal to the within-family heritability (h^2_w):

$$\begin{aligned} 0.1230 &= 1638.92/\ 13324.44 \\ &= Cov(y_{ijkl} - \bar{y}_{\cdot\cdot k\cdot},\ g_{ijkl})\ /\ Var(y_{ijkl} - \bar{y}_{\cdot\cdot k\cdot}) \\ &= [(3/4)\sigma^2_a + (3/4\ tbn)\sigma^2_a]\ /\ Var(y_{ijkl} - \bar{y}_{\cdot\cdot k\cdot}) \\ &= [(3/4)\sigma^2_a + (3/4\ tbn)\sigma^2_a]\ /\ [\sigma^2_{fe}(t-1)/t + \sigma^2_p(tb-1)/tb + \sigma^2_w(tbn-1)/tbn\ ] \\ &\approx h^2_w. \end{aligned}$$

This coefficient approaches h^2_w as t, b, and n increase.

5.9 Multiple Tests of the 'Same' Trait

In previous sections, we have considered coastal Douglas-fir planted in 6 different 10-year-old genetic tests and assumed that the same second moments applied to all 6 tests. In other words, the six tests were all measurements of the same trait, 10-year height. In some cases measurements of the same trait in different environments can be considered different traits. This approach becomes useful when the same trait measured in the different environments has different (or heterogenous) second moments. In the following two examples, we present two situations with heterogeneous moments to illustrate how selection index or BLP combines data in a logical and appealing manner. First (Example 5), we consider a case where observations vary only in their precision or heritability. In forestry, this could arise when some progeny tests are on quite uniform sites, some are on rather variable sites, and/or some are of different design (whether intentionally, or unintentionally as a result of differential mortality). All of these differences could contribute to substantial differences in the variances associated with different observations. Next (Example 6), we consider a slightly more complex case, where observations differ not only in precision, but also in correlation with the genetic value being predicted.

5.9.1 Example 5: Multiple observations differing in precision

Consider again the 10-year-old Douglas-fir tests with genetic parameters as specified for Region B, Table 5.1. In Example 3, a family mean averaged across six tests was used as the unit of observation to predict parental breeding values. Another approach to predicting parental breeding value is to use the six individual test means ($\bar{y}_{i\cdot k\cdot}$) on each parent as observations. The covariance between two test means is

$$\begin{aligned} Cov(\bar{y}_{i\cdot k\cdot},\bar{y}_{i'\cdot k\cdot}) &= Cov(\mu + f_k + fe_{ik} + p_{i\cdot k}/b + e_{i\cdot k\cdot}/bn, \\ &\quad \mu + f_k + fe_{i'k} + p_{i'\cdot k}/b + e_{i'\cdot k\cdot}/bn) \\ &= Cov(f_k,f_k) \\ &= \sigma^2_f \end{aligned} \qquad \textbf{Eq. 5.43}$$

For any one test mean, Eqs. 5.20 and 5.18 can be used to calculate $Var(\bar{y}_{i\cdot k\cdot}) = 2111.33$, and $Cov(\bar{y}_{i\cdot k\cdot},g_k) = 1108$, and Eq. 5.43 can be used to calculate $Cov(\bar{y}_{i\cdot k\cdot},\bar{y}_{i'\cdot k\cdot}) = 554$. The vectors **y**, $\boldsymbol{\alpha}$, and **c** are 6 x 1 vectors, and **V** is a 6 x 6 matrix as below:

$$\mathbf{y} = \begin{bmatrix} \bar{y}_{1 \cdot k \cdot} \\ \bar{y}_{2 \cdot k \cdot} \\ \bar{y}_{3 \cdot k \cdot} \\ \bar{y}_{4 \cdot k \cdot} \\ \bar{y}_{5 \cdot k \cdot} \\ \bar{y}_{6 \cdot k \cdot} \end{bmatrix} \qquad \boldsymbol{\alpha} = E(\mathbf{y}) = \begin{bmatrix} \mu + E_1 \\ \mu + E_2 \\ \mu + E_3 \\ \mu + E_4 \\ \mu + E_5 \\ \mu + E_6 \end{bmatrix}$$

$$\mathbf{c} = \begin{bmatrix} 1108 \\ 1108 \\ 1108 \\ 1108 \\ 1108 \\ 1108 \end{bmatrix} \qquad \mathbf{V} = \begin{bmatrix} 2111.3 & 554 & 554 & 554 & 554 & 554 \\ 554 & 2111.3 & 554 & 554 & 554 & 554 \\ 554 & 554 & 2111.3 & 554 & 554 & 554 \\ 554 & 554 & 554 & 2111.3 & 554 & 554 \\ 554 & 554 & 554 & 554 & 2111.3 & 554 \\ 554 & 554 & 554 & 554 & 554 & 2111.3 \end{bmatrix}$$

Then **c** and **V** are used to calculate the vector of coefficients

$$\boldsymbol{\beta}' = [0.227 \quad 0.227 \quad 0.227 \quad 0.227 \quad 0.227 \quad 0.227],$$

and $\hat{g}_k$ is predicted with the equation

$$\hat{g}_k = 0.227\,(\bar{y}_{1 \cdot k \cdot} - \alpha_1) + 0.227\,(\bar{y}_{2 \cdot k \cdot} - \alpha_2) + 0.227\,(\bar{y}_{3 \cdot k \cdot} - \alpha_3) + 0.227\,(\bar{y}_{4 \cdot k \cdot} - \alpha_4) + 0.227\,(\bar{y}_{5 \cdot k \cdot} - \alpha_5) + 0.227\,(\bar{y}_{6 \cdot k \cdot} - \alpha_6)$$

where $\alpha_i = E(\bar{y}_{i \cdot k \cdot})$. Note that the family mean from every test gets the same amount of weight in the prediction of the breeding value, which is reasonable since each test is of the same design, has the same genetic parameters, is equally precise and equally correlated with the true breeding value. In Example 3 with the same numbers of tests and genetic parameters, we used the family mean across all tests ($\bar{y}_{\cdot\cdot k\cdot}$) to predict the breeding value of the parent, and the equation developed was

$$\hat{g}_k = 1.3619\ (\bar{y}_{\cdot\cdot k\cdot} - \alpha).$$

It can be demonstrated that these two equations will give identical predictions of breeding values of identical precision. Multiplying the six deviations from each test ($\bar{y}_{i\cdot k\cdot} - \alpha_i$) by 0.227 and summing these is equivalent to multiplying the mean of the six test deviations ($\bar{y}_{\cdot\cdot k\cdot} - \alpha$) by 1.3619 (note that 6 x 0.227 = 1.3619). Thus, when all observations have the same genetic parameters (i.e., homogeneous moments) the observations can either be used in the data vector **y** separately in a multiple trait approach, or aggregated into an overall mean resulting in a single element in **y** in a single trait approach; both give identical results. However, if observations have different genetic parameters (heterogeneous moments), only the multiple trait approach is optimum.

Assume that three of the six Douglas-fir tests (say tests 1, 2, and 3) have genetic parameters as presented for Region B, Table 5.1. The other three tests (4, 5, and 6) are planted on environmentally variable sites, and have substantially lower heritabilities as a result (σ^2_p = 815, σ^2_w = 25543, h^2 = 0.08). In these three tests, the variance of a test mean is $Var(\bar{y}_{i\cdot k\cdot})$ = 3278.33 from Eq. 5.20, compared to 2111.33 for the three more precise tests. The covariance between the observed test mean on these three sites and the true breeding value of the parent ($Cov(\bar{y}_{i\cdot k\cdot}, g_k)$ = 1108 from Eq. 5.18) contains only additive genetic covariance, and we assume this is not affected by the increase in environmental error. Similarly, the covariance between a test mean from an imprecise site with a test mean from a precise site is also not affected, thus $\text{Cov}(\bar{y}_{i \cdot k \cdot}, \bar{y}_{i' \cdot k \cdot}) = 554$ from Eq. 5.43 for every pair of tests. Thus, the only change from the homogeneous multiple trait situation above is that the variance of family means from the three imprecise tests 4, 5, and 6 (the lower three diagonal elements of **V**) increased from 2111.33 to 3278.33. The vector of coefficients $\boldsymbol{\beta}' = \mathbf{c}'\mathbf{V}^{-1}$ is

$$\boldsymbol{\beta}' = [0.2657 \quad 0.2657 \quad 0.2657 \quad 0.1519 \quad 0.1519 \quad 0.1519].$$

Then $\hat{g}_k$ is predicted with the equation

$$\hat{g}_k = 0.2657\,(\bar{y}_{1 \cdot k \cdot} - \alpha_1) + 0.2657\,(\bar{y}_{2 \cdot k \cdot} - \alpha_2) + 0.2657\,(\bar{y}_{3 \cdot k \cdot} - \alpha_3) \\ + 0.1519\,(\bar{y}_{4 \cdot k \cdot} - \alpha_4) + 0.1519\,(\bar{y}_{5 \cdot k \cdot} - \alpha_5) + 0.1519\,(\bar{y}_{6 \cdot k \cdot} - \alpha_6)\ .$$

The three family means from the more precise tests get more weight in predicting the breeding value than do the family means from the imprecise tests. Since $\text{Var}(\bar{y}_{i \cdot k \cdot})$ is higher in tests 4, 5, and 6, those family means contain more error, and do not provide as reliable information about genetic effects as do the means from tests 1, 2, and 3. Therefore, less weight is placed on them when predicting the genetic value.

If all observations on all candidates have homogeneous moments (e.g., mass selection or family selection discussed earlier), selection based on either phenotypic or predicted genetic values will result in the same selections and same genetic gain. It is important to note that in any selection situation with heterogeneous moments, however, optimal ranking of candidates cannot be done on the basis of phenotypic values alone. To illustrate this, assume there are two families (D and E) which had the same overall phenotypic family means across the six tests (i.e., $\bar{y}_{\cdot\cdot D\cdot} = \bar{y}_{\cdot\cdot E\cdot}$), but performed differently in different tests (i.e., $\bar{y}_{i\cdot D\cdot} \neq \bar{y}_{i\cdot E\cdot}$ for all i). If family D had its three largest means in tests 1, 2, and 3, while family E had its three largest in tests 4, 5, and 6, the predicted breeding value for parent D could be substantially larger than that for parent E (since D performed best in the more precise tests less affected by error). So, the phenotypic ranks of the two parents (based on the overall family means, which are equal) could be significantly different than ranks based on predicted breeding values. This could easily lead to different selection decisions when using phenotypic ranks compared to those based on predicted breeding values.

Finally, compare the variance of the predicted breeding values and correlation between true and predicted breeding values for the homogeneous moment case (Example 3) and the heterogeneous moment case (Example 5) where test 4, 5, and 6 are less precise. Eqs. 5.6 and 5.8 can be used to calculate $\text{Var}(\hat{g}) = 1509$ and $\text{Corr}(g,\hat{g}) = 0.82$ versus $\text{Var}(\hat{g}) = 1388$ and $\text{Corr}(g,\hat{g}) = 0.79$ for Examples 2 and 5, respectively. When the observed data are less precise (i.e., less heritable) there are two results: 1) predicted genetic values are regressed more toward zero, hence have a lower variance, and 2) correlation between true and predicted genetic values decreases. These are the same effects observed in the simplest case of mass selection (Figure 5.1).

5.9.2 Example 6: Multiple observations differing in precision and correlation with the genetic value

In Example 5, multiple observations of the 'same' trait, height at 10 years, were treated as different traits because of differences in precision. The results illustrate that the BLP approach combines data into a single predicted genetic value in an intuitively appealing manner. This section illustrates a slightly more complex case where the measured traits vary not only in precision, but also in degree of correlation with the genetic value being predicted.

Assume that there are three 10-year-old Douglas-fir progeny tests (1, 2, and 3), all including the same families, which are established in three distinct types of environments or sites (Regions A, B and C), for which genetic parameters are given in Table 5.1. The three regions may be different elevations, latitudes, soil types, etc. Heritabilities in the regions differ, as do $r_{g:R,R'}$, the genetic correlations between height growth in different pairs of regions (Table 5.1). However, $r_{g:R,R'}$ for different pairs of regions is not zero; thus, for example, height growth of a family in Region A provides information about the breeding value of a parent not only in Region A, but also in Regions B and C. Therefore, the observed family means in all three tests (i.e., the three $\bar{y}_{i\cdot k\cdot}$) can be used to predict the breeding value of a parent for each of the three regions (i.e., g_k^A, g_k^B, g_k^C). The covariance between family means measured in different regions is

$$\begin{aligned} \mathrm{Cov}(\bar{y}^{R}_{i\cdot k\cdot},\bar{y}^{R'}_{i'\cdot k\cdot}) &= \mathrm{Cov}(\mu^{R} + f^{R}_{k} + fe^{R}_{ik} + p^{R}_{i\cdot k}/b + e^{R}_{i\cdot k\cdot}/bn, \\ &\qquad \mu^{R'} + f^{R'}_{k} + fe^{R'}_{i'k} + p^{R'}_{i'\cdot k}/b + e^{R'}_{i'\cdot k\cdot}/bn) \\ &= \mathrm{Cov}(f^{R}_{k},f^{R'}_{k}) \qquad\qquad \textbf{Eq. 5.44} \\ &= 1/4\ \mathrm{Cov}(g^{R}_{k},g^{R'}_{k}) \end{aligned}$$

where the R superscripts indicate Region, R = A, B or C, and the linear model for y^R_{ijkl} is Eq. 5.13 for all Regions.

The covariance between a family mean measured in one region with the breeding value in another region ($Cov(\bar{y}^R_{i\cdot k\cdot}, g^{R'}_k)$)is:

$$\begin{aligned} Cov(\bar{y}^R_{i\cdot k\cdot}, g^{R'}_k) &= Cov(\mu^R + f^R_k + fe^R_{ik} + p^R_{i\cdot k}/b + e^R_{i\cdot k\cdot}/bn,\ g^{R'}_k) \\ &= Cov(1/2\ g^R_k, g^{R'}_k) \\ &= 1/2\ Cov(g^R_k, g^{R'}_k) \end{aligned} \qquad \textbf{Eq. 5.45}$$

Equations 5.20, 5.44, and 5.45 are used to estimate the second moments in **V** and **c**:

$$\begin{aligned} Var(\bar{y}^A_{1\cdot k}) &= 1646.67 & Cov(\bar{y}^A_{1\cdot k\cdot}, \bar{y}^B_{2\cdot k\cdot}) &= 332.4 \\ Var(\bar{y}^B_{2\cdot k}) &= 2111.33 & Cov(\bar{y}^A_{1\cdot k\cdot}, \bar{y}^C_{3\cdot k\cdot}) &= 221.6 \\ Var(\bar{y}^C_{3\cdot k}) &= 2111.33 & Cov(\bar{y}^B_{2\cdot k\cdot}, \bar{y}^C_{3\cdot k\cdot}) &= 387.8 \end{aligned}$$

$$\begin{aligned} Cov(\bar{y}^A_{1\cdot k\cdot}, g^A_k) &= 1108 & Cov(\bar{y}^A_{1\cdot k\cdot}, g^B_k) &= 664.8 & Cov(\bar{y}^A_{1\cdot k\cdot}, g^C_k) &= 443.2 \\ Cov(\bar{y}^B_{2\cdot k\cdot}, g^A_k) &= 664.8 & Cov(\bar{y}^B_{2\cdot k\cdot}, g^B_k) &= 1108 & Cov(\bar{y}^B_{2\cdot k\cdot}, g^C_k) &= 775.6 \\ Cov(\bar{y}^C_{3\cdot k\cdot}, g^A_k) &= 443.2 & Cov(\bar{y}^C_{3\cdot k\cdot}, g^B_k) &= 775.6 & Cov(\bar{y}^C_{3\cdot k\cdot}, g^C_k) &= 1108 \end{aligned}$$

Then the form of the necessary vectors and matrices **y**, $\boldsymbol{\alpha}$, **V**, $\mathbf{c}^A$, $\mathbf{c}^B$, $\mathbf{c}^C$ are below, where $\mathbf{c}^A$, $\mathbf{c}^B$, and $\mathbf{c}^C$ are the vectors used when predicting breeding values for Regions A, B, and C, respectively:

$$\mathbf{y} = \begin{bmatrix} \bar{y}^A{}_{1\cdot k\cdot} \\ \bar{y}^B{}_{2\cdot k\cdot} \\ \bar{y}^C{}_{3\cdot k\cdot} \end{bmatrix} \qquad \boldsymbol{\alpha} = E(\mathbf{y}) = \begin{bmatrix} \mu + E_1 \\ \mu + E_2 \\ \mu + E_3 \end{bmatrix}$$

$$\mathbf{V} = \begin{bmatrix} 1646.67 & 332.4 & 221.6 \\ 332.4 & 2111.33 & 387.6 \\ 221.6 & 387.6 & 2111.33 \end{bmatrix}$$

$$\mathbf{c}^A = \begin{bmatrix} 1108 \\ 664.8 \\ 443.2 \end{bmatrix} \quad \mathbf{c}^B = \begin{bmatrix} 664.8 \\ 1108 \\ 775.6 \end{bmatrix} \quad \mathbf{c}^C = \begin{bmatrix} 1108 \\ 664.8 \\ 443.2 \end{bmatrix}$$

These vectors and matrices are used to calculate three β vectors; each one contains three coefficients for predicting the parental breeding values for a different region from the same test data from all three regions. For this example, the equations to predict breeding values for each of the three regions are:

$$\hat{g}^A = 0.6184\,(\bar{y}^A{}_{1\cdot k\cdot} - \mu - E_1) + 0.1975\,(\bar{y}^B{}_{2\cdot k\cdot} - \mu - E_2) + 0.1087\,(\bar{y}^C{}_{3\cdot k\cdot} - \mu - E_3),$$

$$\hat{g}^B = 0.2816\,(\bar{y}^A{}_{1\cdot k\cdot} - \mu - E_1) + 0.4330\,(\bar{y}^B{}_{2\cdot k\cdot} - \mu - E_2) + 0.2583\,(\bar{y}^C{}_{3\cdot k\cdot} - \mu - E_3),$$

$$\hat{g}^C = 0.1550\,(\bar{y}^A{}_{1\cdot k\cdot} - \mu - E_1) + 0.2583\,(\bar{y}^B{}_{2\cdot k\cdot} - \mu - E_2) + 0.4611\,(\bar{y}^C{}_{3\cdot k\cdot} - \mu - E_3).$$

From these three equations, a number of points can be observed that illustrate how BLP combines multiple traits to predict genetic worth:

1) For predicted breeding values for all three regions, the family mean observed in that region gets the largest weight. This occurs primarily because the mean within a region ($\bar{y}^R_{i\cdot k\cdot}$) has the highest correlation with parental breeding value (g^R) in that region. (Note that $r_{g:R,R'}$ by definition equals 1.0).

2) In the equation for predicting g^A, $\bar{y}^A_{1\cdot k\cdot}$ gets substantially more weight than does $\bar{y}^B_{2\cdot k\cdot}$ in predicting g^B, or $\bar{y}^C_{3\cdot k\cdot}$ in predicting g^C. This is because $\bar{y}^A_{1\cdot k\cdot}$ has the highest heritability of the three observations (compare h^2 for the three regions in Table 5.1).

3) The prediction equation for g^B places relatively more weight on the observed mean from the other two regions than do the equations for g^A and g^C. The genetic correlations between breeding values in different regions (Table 5.1) illustrate why this occurs. Heuristically, the observed means $\bar{y}^A_{1\cdot k\cdot}$ and $\bar{y}^C_{3\cdot k\cdot}$ provide some information about breeding values in Region B. In comparison, for breeding values in Regions A and C, the observed means in the other two regions provide less information (they are less correlated with g^A or g^C), thus relatively higher weights are placed on $\bar{y}^A_{1\cdot k\cdot}$ and $\bar{y}^C_{3\cdot k\cdot}$, respectively.

4) The observations $\bar{y}^B_{2\cdot k\cdot}$ and $\bar{y}^C_{3\cdot k\cdot}$ have the same heritability, but $\bar{y}^B_{2\cdot k\cdot}$ gets a higher weight in predicting g^A because it is more strongly correlated ($r_{g:A,B} = 0.6$, $r_{g:A,C} = 0.4$).

Different traits always have heterogeneous moments (almost by definition). Very complex combinations of different traits may confront the breeder. For example, tests may be of different ages, (say 3-, 5-, and 10-year measurements), or data may be available on distinct traits (e.g., height, straightness, specific gravity) or from different relatives (ortet, ramet, half-sib, full-sib, etc.), or various combinations of these. These two examples illustrate that as long as **V** and **c** are specified correctly, the multiple trait BLP or selection index approach develops a linear combination of the data that implicitly accounts for 1) the precision or heritability of every observation being used, and 2) its degree of correlation with all other observations and the genetic value being predicted. Although it has not been illustrated with the examples in this chapter, the approach also accounts for any differences in variances or covariances attributable to effects of scale. Thus, the approach to predicting genetic values and genetic gain in complex multiple trait situations is conceptually the same as in the simplest case of direct mass selection.

5.10 Summary

All selection should be based on predicted genetic worth ($\hat{g}$). If a predicted genetic worth is calculated with the general BLP or selection index approach such that

$$\begin{aligned}\hat{g} &= \tau + [\text{Cov(observed, predictand)/Var(observed)}]\ (y-\alpha)\\ &= \tau + [\text{Cov}(g,y)/\text{Var}(y)]\ (y-\alpha),\end{aligned}$$

then $\hat{g}$ is optimum in the sense that it minimizes the squared error of prediction, i.e., $E(g-\hat{g})^2$ within the class of linear functions. Heuristically, $\hat{g}$ is calculated by regressing or shrinking observed deviations according to 1) the *effective heritability* of the observed data, or, in other words, the precision with which measured traits reflect genetic effects, and 2) the correlation between the genetic effects of the observed data and the predictand.

Predicted genetic gain is the average $\hat{g}$ of the individuals selected. If the breeder has equal information on all candidates, then selection on the observed data (e.g., an individual tree measurement or a family mean) results in the same selections and the same genetic gain achieved as if selection were done on $\hat{g}$. By equal information, we mean that the same genetic parameters (h^2, σ^2_y, Corr(g,y)) are applicable to all measurements on all candidates and that all candidates are equally tested in every sense. In forestry breeding programs of any size, having equal information in this sense may be a relatively rare occurrence. In any kind of multiple trait situation, (i.e., any case with unequal information: progeny tests of different ages, different genetic parameters, different numbers of measurements, etc.), a selection index (BLP) approach is needed to make correct selections, and maximize and predict gain. Any other approach is less than optimal, although the loss in efficiency depends how severely assumptions are violated, i.e., how unequal, unbalanced etc. are the data.

The derivation of all formulae for prediction of genetic worth and genetic gain consider only linear combinations of the observed data, and assume that second moments (variances and covariances) are known. Second moments are never truly known, but if the estimates used are reasonably good, the ranking of parents will be close to optimum, therefore selections will generally be correct and the gain efficiency (the ratio of actual gain achieved versus gain possible if true parameters were known) will be high.

Tree improvement programs are becoming more and more complex with large amounts of data already collected. Also, an array of different types of data may be measured on future trees in advanced generations. The diverse data available to the breeder may include measurements from a variety of field genetic tests for disease, growth, wood quality, etc., data from early selection tests in greenhouses or nurseries, decades of progeny test data on ancestors, and molecular or physiological measurements which may be correlated with traits of interest. A selection index or best linear prediction approach allows a breeder to consider all information available when ranking candidates for selection in a system which is objective, consistent, and can be improved as better estimates of second moments become available.

Literature Cited and Selected References

Becker, W.A. 1975. *Manual of Quantitative Genetics.* Washington State University Press. Pullman, WA. 170pp.

Binet, F.E. 1965. On the construction of an index for indirect selection. *Biometrics* **21**:291-299.

Bridgwater, F.E. and A.E. Squillace. 1986. Selection indexes for forest trees. P. 17-20 in *Advanced Generation Breeding of Forest Trees*. Southern Coop. Series Bull. 309, Louis. Ag. Exp. Stn., Baton Rouge, LA.

Bridgwater, F.E. and R.W. Stonecypher. 1979. Index selection for volume and straightness in a loblolly pine population. P. 132-139 in *Proc. 15th. South. For. Tree Improve. Conf.*, June 19-21, Miss State, MS.

Bulmer, M.G. 1985. *The Mathematical Theory of Quantitative Genetics*. Clarendon Press, Oxford, GB. 254p.

Burdon, R.D. 1979. Generalisation of multi-trait selection indices using information from several sites. *New Zealand J. For. Sci.* **9**:145-152.

Burdon, R.D. 1982. Selection indices using information from multiple sources for the single-trait case. *Silvae Genet.* **31**:81-85.

Cotterill, P. and N. Jackson. 1981. Index selection with restrictions in tree breeding. *Silvae Genet.* **30**:106-108.

Cotterill, P.P., R.L. Correll and R. Boardman. 1983. Methods of estimating the average performance of families across incomplete open-pollinated progeny tests. *Silvae Genet*. **32**:28-32.

Falconer, D.S. 1965. The inheritance of liability to certain diseases, estimated from the incidence among relatives. *Ann. Hum. Genet.* **29**:51-76.

Falconer, D.S. 1981. *Introduction to Quantitative Genetics*. Longman & Co., New York, NY. 340p.

Fisher, R.A. and F. Yates. 1953. *Statistical tables for biological, agricultural, and medical research*. Hafner Publishing Co., NY. 126pp.

Harris, D.L. 1963. The influence of errors of parameter estimation upon index selection. P. 491-500 in *Statistical Genetics and Plant Breeding*, Nat. Acad. Sciences, Nat. Research Council, Washington, DC.

Harris, D.L. 1964. Expected and predicted progress from index selection involving estimates of the population parameters. *Biometrics* **20**:46-72.

Hazel, L.N. 1943. The genetic basis for constructing selection indices. *Genetics* **38**:476-490.

Henderson, C.R 1975. Best linear unbiased estimation and prediction under a selection model. *Biometrics* **31**:423-447.

Henderson, C.R 1977. Best linear unbiased prediction of breeding values not in the model for records. *J. Dairy Sci.* **60**:783-787.

Henderson, C.R. 1963. Selection index and expected genetic advance. P. 141-163 in *Statistical Genetics and Plant Breeding*, W.D. Hanson and H.F. Robinson (eds.). NAS-NRC Pub. No. 982, Washington, DC.

Henderson, C.R. 1973. Sire evaluation and genetic trends. P. 10-41 in *Animal Breeding and Genetics Symposium in Honor of J. Lush*, Animal Sci. Assoc. Amer., Champaign, Ill.

Henderson, C.R. 1984. *Applications of Linear Models in Animal Breeding.* University of Guelph, Guelph, Ontario. 462p.

Namkoong, G. 1979. *Introduction to Quantitative Genetics in Forestry.* USDA Forest Service Technical Bulletin No. 1588. Washington, D.C. 342p.

Namkoong, G., E.B. Snyder and R.W. Stonecypher. 1966. Heritability and gain concepts for evaluating breeding systems such as seedling orchards. *Silvae Genet.* **15**:76-84.

Neter, J., and W. Wasserman. 1974. *Applied Linear Statistical Models.* Richard D. Irwin and Co., Homewood, IL. 842pp.

Noble, B., and J.W. Daniel. 1977. *Applied Linear Models.* Prentice-Hall, Inc., Englewood Cliffs, NJ. 477pp.

Sales, J. and W.G. Hill. 1976. Effect of sampling errors on efficiency of selection indices. *Anim. Prod.* **22**:1-17.

Searle, S.R. 1961. Phenotypic, genetic and environmental correlations. *Biometrics* **17**:474-480.

Searle, S.R. 1966. *Matrix Algebra for the Biological Sciences.* John Wiley and Sons, NY. 296pp.

Shelbourne, C.J.A 1969. *Tree breeding methods.* For. Res. Inst. Tech. Pap. No. 55, New Zealand For. Serv., Wellington, New Zealand.

Shelbourne, C.J.A. and C.B. Low. 1980. Multi-trait index selection and associated genetic gains of Pinus radiata progenies at five sites. *N. Z. J. For. Sci.* **10**:307-324.

Smith, H. 1936. A discriminant function for plant selection. *Ann. Eug.* 7:240-250.

Sorensen, F.C. and T. L. White. 1988. Effect of natural inbreeding on variance structure in tests of wind-pollinated Douglas-fir progenies. *For. Sci.* **34**:102-118.

Squillace, A. E. 1974. Average genetic correlations among offspring from open-pollinated forest trees. *Silvae Genetica* **23**:149-156.

Thompson, R. 1979. Sire evaluation. *Biometrics* **35**:339-353.

Van Vleck, L.D., E.J. Pollack, and E.A. Branford. 1987. *Genetics for the Animal Sciences*. W.H. Freeman and Co., NY. 391pp.

White, T.L. and G.R. Hodge. 1989. *Predicting Breeding Values with Applications in Forest Tree Improvement*. Kluwer Academic Pub., Dordrecht. 367pp.

White, T.L., G.R. Hodge and M.A. Delorenzo. 1986. Best linear prediction of breeding values in forest tree improvement. P. 99-122 in *Proc. Statistical Considerations in Genetic Testing of Forest Trees*. South.Coop.Series Bull.324, Univ. of Florida, Gainesville, FL, 32611.

Chapter 6
Computational Methods

R. W. Stonecypher[1]

Analyses and interpretation of field and laboratory experiments are facilitated by the wide acceptance and use of a statistical technique known as the analysis of variance, often called "ANOVA". The technique, developed by R. A. Fisher more than seven decades ago, has stood the test of time in providing an extremely powerful analytical tool, not only for the interpretation of results, but also as a catalyst for using sound experimental designs, particularly in agronomic and forestry field studies.

Within the last two decades, very powerful, cost-effective, computational tools have become readily available to practically all forestry researchers. Armed with these tools, forest tree breeders can carry out complex ANOVA'S of very large and even unbalanced (non-orthogonal) data sets that would have been all but impossible twenty years ago. While the availability of these tools is a plus, the researcher should not apply them indiscriminately, and above all, avoid "Driving tacks with pile drivers".

---

[1]*Research Advisor, Tree Improvement, Weyerhaeuser Company, Centralia, Washington*

L. Fins et al. (eds.), Handbook of Quantitative Forest Genetics, 195–228.

6.1 The Analysis of Variance (ANOVA)

This chapter briefly discusses basic principles of ANOVA, and presents worked examples from real-world forest genetics studies. I will use SAS®[2] as the basic computational tool in developing the examples, since this system is widely used and generally available to forest genetics researchers. The use of SAS® procedures, run stacks and outputs also will make my task easier, and will serve to demonstrate some "tricks and traps" that the researcher may relate to their research data. I am purposely avoiding an extensive sortie into statistical theory, leaving it to the reader to reconnoiter this area as deemed necessary.

6.1.1 Classes of problems solvable

The discussions in Sections 6.1.1 through 6.1.3.1 are based largely on a very excellent presentation of ANOVA by Eisenhart, 1947.

The ANOVA procedure can be used to address two types of problems:

1. Estimating and testing to infer the existence of true differences among means.
2. Estimating and testing to infer the existence of components of variance.

Problem 1 above requires a Type I Model ("Fixed"), and problem 2, a Type II Model ("Random"). In trying to determine which model applies to a given experiment, we might ask the following questions:

---

[2]*SAS® is the registered trademark of SAS Institute, Inc.*

1. Are the conclusions confined to the things actually studied; to the immediate sources of these things; or extended to apply to more general populations?

2. In complete repetitions of the experiment would the same things be studied again; would new samples be drawn from the same sources; or would new samples be drawn from the general populations?

A rule of thumb for most forest genetics studies is that when the objective of the particular experiment is to estimate components of variance, and the assumptions of random effects are met, the model is Type II. If the objective of the study is to estimate differences among means (effects), the model is either Type I, or more often, a mixture of I and II. It should be noted, however, that the two types of objectives may not always be mutually exclusive. If the model is clearly Type II, there are no complications in relation to tests of significance or the generalization of the results for the sources of variance. If the model is mixed, the significance tests may be more complicated, and the generalization of the conclusions may be circumscribed. I will attempt to clarify the above guidelines, and their relevance to developing tests of significance with an example later.

6.1.2 A numerical example

The computational methods used to develop an ANOVA table are the same for Model I and Model II. This is true because the decomposition of the total sum of squared deviations is an algebraic identity. It should be emphasized that the arithmetic of ANOVA is valid whatever the meaning of the numbers involved.

The contrived numerical example used in Boxes 6.1A and 6.1B demonstrates the algebraic identities of ANOVA, and I will leave it to readers to verify them for their satisfaction. I have also included the output from a SAS® run of PROC ANOVA using the same data set. Note that the practical formulas would normally be used to develop an ANOVA table if done manually.

6.1.3 Assumptions underlying the use of ANOVA in statistical inference

As stated earlier, the arithmetic procedures used for subdividing a data set by decomposition of the total sums of squares require no validating assumptions. If, however, one wishes to make statistical inferences about the data using a test statistic, the usual case, then certain assumptions must be fulfilled if the inferences are to be valid.

Assumptions for Model I:

1. The observations are observed values of random variables distributed about true mean values m_{ij}.
2. The m_{ij} are simple additive functions of the marginal (row and column), and general means.
3. The observations have common variances and zero covariances.
4. The observations are normally distributed.

Box 6.1A. Basic analysis of variance (ANOVA) formulae and calculations. From Eisenhart, 1947.

ROWS	COLUMNS 1	2	3	4	5	Xi.	$\bar{X}$i.
1	23	25	28	35	15	126	25.2
2	30	38	44	53	22	187	37.4
3	15	17	23	30	14	99	19.8
4	10	12	17	25	11	75	15.0
X.j	78	92	112	143	62	X..= 487	
$\bar{X}$.j	19.50	23.00	28.00	35.75	15.50	$\bar{X}$..	24.35

$$(Xij - \bar{X}..) = (\bar{X}i. - \bar{X}..) + (\bar{X}.j - \bar{X}..) + (Xij - \bar{X}i. - \bar{X}.j + \bar{X}..) \quad [1]$$

$\Sigma(X - \bar{X})^2 = \Sigma X^2 - (\Sigma X)^2/N$ [2]; N = Total Observations.

Squaring [1]:

$$\Sigma(Xij - \bar{X}..)^2 = \Sigma(\bar{X}i. - \bar{X}..)^2 + \Sigma(\bar{X}.j - \bar{X}..)^2 + \Sigma(Xij - \bar{X}i. - \bar{X}.j + \bar{X}..)^2$$

PRACTICAL FORMULAS [2]	Corrected Sums Squares	Mean Squares
TOT CSS = $\Sigma\Sigma Xij^2 - X^2../20$	2520.55	
ROW CSS = $\Sigma X^2 i./5 - X^2../20$	1395.75	465.25
COL CSS = $\Sigma X^2.j/4 - X^2../20$	987.80	246.95
RXC CSS = TOT CSS - ROW CSS - COL CSS	137.00	11.42

CSS = Corrected Sums of Squares.

Box 6.1B. Output from a SAS® run of data set in Box 6.1A.

Analysis of Variance Procedure
Class Level Information

Class	Levels	Values
ROW	4	1 2 3 4
COL	5	1 2 3 4 5

Number of observations in data set = 20

Analysis of Variance Procedure

Dependent Variable: X

Source	DF	Sum of Squares	Mean Squares	F Value	Pr > F
Model	7	2383.55	340.50	29.83	0.0001
Error	12	137.00	11.41		
Corrected Total	19	2520.55			

R-Square	C.V.	Root MSE	X Mean
0.945647	13.87	3.37	24.35

Source	DF	Anova SS	Mean Square	F Value	Pr >F
ROW	3	1395.75	465.25	40.75	0.0001
COL	4	987.80	246.95	21.63	0.0001

If the above assumptions are satisfied, procedures for estimating and testing inferences about non-zero differences among population means (Model I), and existence of components of variance (Model II), are valid. Note, however, that it is not possible to make exact tests of the absolute values or derive exact confidence limits for the components of variance.

6.1.3.1 Consequences and possible rectifications when the assumptions are not satisfied

1. *Nonrandomness*. Results in biased estimates of true effects. Care should be taken to ensure random assignment of effects to experimental design.
2. *Non-additivity of main effects.* F test will be inefficient, and perhaps misleading. Non-additivity is best dealt with by transformation of the original scale of the data measurements.
3. *Heterogeneity of variances.* Moderate levels of heterogeneity are not serious for overall tests of significance, but single degree of freedom comparisons may be highly inaccurate. A weighted means test is appropriate in such cases.
4. *Non-normality.* Generally not serious. Very skewed distributions have a marked effect on the significance level of the F-test. Transformations can be used to make data more normally distributed. If, after transformation, data distribution remains substantially non-normal, nonparametric analytical methods should be considered.

6.2 Analyses of Forest Genetics Experiments: Some Examples

Below are several data sets from forest genetics tests and research plantings from Weyerhaeuser's Tree Improvement Program. These examples should provide useful guides to analyses of large data sets that are common in forest tree experiments. These examples are by no means all inclusive, but represent the most common that I have encountered in my experience over the last 25 years.

6.2.1 Individual tree analyses and variance component estimates

There is often a need to estimate variance components on an individual tree basis in genetic tests. This is so because the unit of selection is the individual, and estimates of genetic and environmental variances and covariances are required for development of selection criteria or gain prediction on an individual tree basis. Since most tree studies are large, and very few have 100 percent survival, analyses on an individual basis may present some difficulties. While computational methods for complete analysis of experiments with missing or "messy" data now exist, these procedures, depending on the size of the data set, the complexity of the study design, and the degree of unbalance, can be **very consumptive** of computer time and memory. The individual tree analyses of some of our genetic tests, although of moderate size, have severely taxed very large mainframe computers.

Analysis of plot means offers an alternative, but does not provide the desired individual based variance component estimates. An optional approach is to estimate the within plot variances, plot means, and harmonic means of number of individuals per plot, separately, and then carry out the complete analysis using plot means. The SAS® run stack (Box 6.2), and the output from the PROC VARCOMP analysis (Box 6.3), illustrate this approach. Box 6.4 contains the results for the individual tree analysis using PROC VARCOMP. Note that, to reduce computing time, the model for the individual tree analysis does not include an estimate of the plot to plot (REP(PLT)*FEM) component of variance.

The estimates of the variance components and their ratios h^2 (heritabilities) in Box 6.5 are very similar for the two types of analyses. The analyses of plot means thus gives estimates desired, and requires much less computational time and computer resources.

6.2.2 Differences among means and analyses of unbalanced data sets
We often face problems of missing cells, as well as missing individuals, in forest genetics experiments. Experiments with large or even moderate numbers of missing cells require special techniques for effective and correct analyses. I will use results from a Weyerhaeuser experiment, designed to examine selected family performance in breeding zones of origin and non-origin, for comparing analytical procedures for dealing with missing cells. In addition, I will use this example to demonstrate the construction of F tests for a mixed model experiment. This study, and the SAS® run stacks used to analyze it, are described in Box 6.6.

An analysis of plot means using SAS® PROC VARCOMP for this Wide Test Study resulted in the ANOVA table and expected mean squares in Box 6.7. Note that the first five effects are fixed (line 13 Box 6.6). The FIXED=5 option specifies the first five effects in the model statement as fixed. The output, at the bottom of Box 6.7, contains variance component estimates from those effects designated random, i. e. not fixed.

Box 6.2. SAS® run stack for PROC VARCOMP using plot means.

```
LIBNAME INDIV 'I:\';
DATA IND;
SET INDIV.CB88;
WHERE FEM GT 0 AND HT6 NE .;
PROC SORT; BY PLT REP FEM;
/* Compute Plot Means and Within Plot SS */;
PROC MEANS NOPRINT MEAN CSS N;BY PLT REP FEM;
VAR HT6;
OUTPUT OUT=FBYR MEAN=FRHT6 CSS=CSSHT6 N=NP;
DATA PLOTM;
SET FBYR;
RN=1/NP;
DF=NP-1;
PROC MEANS NOPRINT  SUM N;VAR CSSHT6 RN DF;
OUTPUT OUT=WPMS SUM=TWSSHT6 TRN TDF N=NPLOTS;
DATA WPMSPRT;
SET WPMS;
WPMSH6=TWSSHT6/TDF;
/* Harmonic Mean of Trees per Plot */;
HM=1/(1/NPLOTS*TRN);
PROC PRINT;VAR TDF HM WPMSH6;
TITLE 'WITHIN PLOT MS FOR  CB SUP SITES';
/* Using  PROC VARCOMP TYPE I.
I am assuming a Type II Model. */;
PROC VARCOMP DATA=FBYR  METHOD=TYPE1;
CLASS PLT REP FEM;
MODEL FRHT6=PLT REP(PLT) FEM PLT*FEM;
TITLE 'CB SUP SITES VARCOMP TYPE1 PLOT MEAN BASIS';

OUTPUT FROM WITHIN PLOT MEAN SQUARES ANALYSIS:
                           WITHIN PLOT MS FOR  CB SUP SITES
                           OBS       TDF       HM        WPMSH6
                            1        4053      3.45        0.29477
```

Box 6.3. Output from plot mean PROC VARCOMP analysis.

```
         CB SUP SITES VARCOMP TYPE1 PLOT MEAN BASIS
          Variance Components Estimation Procedure
                   Class Level Information

Class     Levels     Values
PLT         16       25 26 27 28 29 30 31 32 33 34 35 36 37
                     38 39 40
REP          6       1 2 3 4 5 6
FEM         16       1 9 16 20 31 47 48 53 60 61 68 75 76
                     88 98 102
           Number of observations in data set = 1536

       Variance Components Estimation Procedure
 Dependent Variable: FRHT6
Source                DF     Type I SS      Type I MS
PLT                   15     1065.0149        71.0009
REP(PLT)              80       71.0591         0.8882
FEM                   15       40.4025         2.6935
PLT*FEM              225       30.8455         0.1370
Error               1200      107.8223         0.0898
Corrected Total     1535     1315.1445

Source        Expected Mean Squares
PLT          Var(Error) +  6 Var(PLT*FEM)
             + 16 Var(REP(PLT))+ 96 Var(PLT)
REP(PLT)     Var(Error) + 16 Var(REP(PLT))
FEM          Var(Error) +  6 Var(PLT*FEM) + 96 Var(FEM)
PLT*FEM      Var(Error) +  6 Var(PLT*FEM)
Error        Var(Error)

Variance Component          Estimate
Var(PLT)                     0.7298
Var(REP(PLT))                0.0498
Var(FEM)                     0.0266
Var(PLT*FEM)                 0.0078
Var(Error)                   0.0898
```

Box 6.4. Output from individual tree PROC VARCOMP analysis.

```
          CB SUP SITES VARCOMP TYPE1 INDIVIDUAL BASIS

             Variance Components Estimation Procedure

                     Class Level Information

Class      Levels     Values
PLT          16       25 26 27 28 29 30 31 32 33 34 35 36 37
                      38 39 40
REP           6        1 2 3 4 5 6
FEM          16        1 9 16 20 31 47 48 53 60 61 68 75 76
                      88 98 102

          Number of observations in data set = 5589

            Variance Components Estimation Procedure

Dependent Variable:  HT6           HT in m


Source                 DF       Type I SS       Type I MS
PLT                    15       3863.0804        257.5386
REP(PLT)               80        257.4757          3.2184
FEM                    15        147.2975          9.8198
PLT*FEM               225        108.6205          0.4827
Error                5253       1562.2744          0.2974
Corrected Total      5588       5938.7487
```

Cont'd next page

Box 6.4 cont'd

```
Source                Expected Mean Squares
PLT          Var(Error) + 21.979 Var(PLT*FEM)
             + 0.1042 Var(FEM) + 58.337 Var(REP(PLT))
             + 349.21 Var(PLT)
REP(PLT)     Var(Error) + 0.0774 Var(PLT*FEM)
             + 0.0774 Var(FEM) + 58.192 Var(REP(PLT))
FEM          Var(Error) + 21.988 Var(PLT*FEM)
             + 348.74 Var(FEM)
PLT*FEM      Var(Error) + 21.783 Var(PLT*FEM)
Error        Var(Error)

Variance Component                 Estimate
Var(PLT)                            0.7277
Var(REP(PLT))                       0.0501
Var(FEM)                            0.0267
Var(PLT*FEM)                        0.0085
Var(Error)                          0.2974
```

Box 6.8 has the output from PROC GLM analyses produced by the SAS® run stack at the bottom of Box 6.6. The RANDOM option immediately below the MODEL statement in Box 6.6, specifies that the planting location and replication within planting location effects are random. Box 6.9 contains the TYPE I and III sums of squares, and the F tests. Note that PROC GLM uses the error mean square to test all levels unless specified otherwise. I have developed F tests (F calculation Box 6.10), assuming planting locations and replications are random effects. Note that the PLT(BZ) test is approximate, and adjusted DF should be calculated for the F test. See Snedecor and

Box 6.5. Comparison of variance component estimates from plot mean and individual tree analyses using PROC VARCOMP.

VARIANCE COMPONENTS

	Indiv.	**Means**		
$\sigma^2 f$	0.0268	0.0266	$\sigma^2 f$= Var(FEM)	Box 6.3 & 6.4.
$\sigma^2 fp$	0.0085	0.0078	$\sigma^2 fp$= Var(PLT*FEM)	" " "
Error MS	0.2974	0.0898	Var(Error)	" " "
$\sigma^2 p$	0.0026	0.0046	$\sigma^2 p^1$ = Plot to Plot Variance.	
h^2	0.161	0.159		

Within Plot Mean Square ≡ σ^2= 0.2948
harmonic mean no./plot= 3.46

For Individual Analysis: $\sigma^2 p^1$= 0.2974 - 0.2948

For Plot Mean Analysis: $\sigma^2 p^1$= 0.0898 - .2948/3.46

Cochran, 1976, page 369, for an example. These tests were developed using the expected mean squares from the VARCOMP and GLM analyses. For those of you who have a PC, the construction and calculation of the F tests can be facilitated by use of a spreadsheet program. The F tests also can be made by using the TEST statement in PROC GLM of SAS®. Box 6.11 presents the output from using this approach. It has been my experience that deriving F tests in the more complicated mixed models are best obtained using a two step process by developing expected mean squares (from PROC VARCOMP or GLM), and computing the F values manually or from another GLM run using the TEST statement. I personally prefer the manual approach.

Box 6.6. Design and SAS® run stacks for Weyerhaeuser wide test study.

2 Sources (**SOURCE**): Coos Bay & Springfield.
2 Breeding zones (**BZ**): Coos Bay & Springfield.
3 Planting sites (**PLT**) in each breeding zone.
20 Pollen-mix families (**FEM**) from Coos Bay.
17 Pollen-mix families (**FEM**) from Springfield.
8 Replications (**REP**) in each planting site;
4 trees/replication.
Total trees = 7104

SAS® RUN STACK WIDE TEST STUDY PROC VARCOMP

```
OPTIONS LINESIZE=80 PAGESIZE=60;
* WIDE TEST ANALYSIS OF FAMILY BY REP MEANS;
LIBNAME WIDETEST 'I:\';
TITLE 'WIDE TEST MEANS ANALYSIS';
DATA INDIV;
SET WIDETEST.FAMBYREP;
WHERE FEM GT 0 AND MX3 NE .;
IF FEM LT 400 THEN SOURCE='CB';ELSE SOURCE='SP';
PROC VARCOMP METHOD=TYPE1;
CLASS BZ PLT REP  SOURCE FEM;
MODEL MX3 = BZ SOURCE BZ*SOURCE FEM(SOURCE)
BZ*FEM(SOURCE) PLT(BZ) REP(BZ PLT)
SOURCE*PLT(BZ) PLT*FEM(BZ SOURCE) / FIXED=5;
RUN;
```

SAS® RUN STACK PROC GLM.

```
OPTIONS PS=60 LS=80;
DATA FBYR;
INFILE WTFBYR;
INPUT BZ $ SOURCE $ PLT REP FEM HT5 HT8;
PROC GLM;
TITLE 'WIDE TEST MEANS ANALYSIS PROC GLM';
CLASS BZ PLT REP  SOURCE FEM;
MODEL HT8 = BZ SOURCE BZ*SOURCE FEM(SOURCE)
BZ*FEM(SOURCE)
PLT(BZ) REP(BZ PLT) SOURCE*PLT(BZ) PLT*FEM(BZ SOURCE);
RANDOM PLT(BZ) REP(BZ PLT);
RUN;
```

Box 6.7. Wide test expected mean squares from VARCOMP analysis.

```
Source           Expected Mean Squares
BZ               Var(Error) + 9.192 Var(PLT*FEM(BZ*SOURCE))
                 + 159.75 Var(PLT*SOURCE(BZ))
                 + 39.94 Var(REP(BZ*PLT)) + 319.5
                    Var(PLT(BZ))
                 +Q(BZ,SOURCE,BZ*SOURCE,FEM(SOURCE)
                 ,BZ*FEM(SOURCE))

SOURCE           Var(Error) + 9.1882
                    Var(PLT*FEM(BZ*SOURCE))
                 + 159.75 Var(PLT*SOURCE(BZ))
                 + 0.0025 Var(REP(BZ*PLT))
                 + 0.0017 Var(PLT(BZ)) + Q(SOURCE,BZ*SOURCE
                 ,FEM(SOURCE),BZ*FEM(SOURCE))

BZ*SOURCE        Var(Error) + 9.1882
                    Var(PLT*FEM(BZ*SOURCE))
                 + 159.75 Var(PLT*SOURCE(BZ))
                 + 0.0025 Var(REP(BZ*PLT)) + 0.0017
                    Var(PLT(BZ))
                 + Q(BZ*SOURCE,FEM(SOURCE),BZ*FEM(SOURCE))

FEM(SOURCE)      Var(Error) + 8.6053
                    Var(PLT*FEM(BZ*SOURCE))
                 + 0.0015 Var(PLT*SOURCE(BZ))
                 + 0.0028 Var(REP(BZ*PLT)) + 0.0015
                    Var(PLT(BZ))
                 + Q(FEM(SOURCE),BZ*FEM(SOURCE))

BZ*FEM(SOURCE)   Var(Error) + 8.6048
                    Var(PLT*FEM(BZ*SOURCE))
                 + 0.0015 Var(PLT*SOURCE(BZ))
                 + 0.0028 Var(REP(BZ*PLT)) + 0.0015
                    Var(PLT(BZ))
                 + Q(BZ*FEM(SOURCE))
```

Cont'd next page

Box 6.7 cont'd

```
PLT(BZ)             Var(Error) + 9.1907
                       Var(PLT*FEM(BZ*SOURCE))
                    + 159.74 Var(PLT*SOURCE(BZ))
                    + 39.936 Var(REP(BZ*PLT)) + 319.47
                       Var(PLT(BZ))

REP(BZ*PLT)         Var(Error) + 0.0018
                       Var(PLT*FEM(BZ*SOURCE))
                    + 0.0009 Var(PLT*SOURCE(BZ))
                    + 39.933 Var(REP(BZ*PLT))

PLT*SOURCE(BZ)      Var(Error) + 9.1861
                       Var(PLT*FEM(BZ*SOURCE))
                    + 159.73 Var(PLT*SOURCE(BZ))

PLT*
FEM(BZ*SOURCE)      Var(Error) + 8.6022
                       Var(PLT*FEM(BZ*SOURCE))

Error                          Var(Error)
Variance Component               Estimate
Var(PLT(BZ))                       0.2584
Var(REP(BZ*PLT))                   0.1477
Var(PLT*SOURCE(BZ))                0.0042
Var(PLT*FEM(BZ*SOURCE))           -0.0022
Var(Error)                         0.2512
```

Note the similarity of the SS and particularly the F values for three of the four types of SS (Types I, III & IV) which can be calculated by PROC GLM. In cases where there are missing cells in the data, Type III & IV SS may not agree, and F values and tests of significance using Type IV SS are more appropriate.

Box 6.8. Expected mean squares wide test PROC GLM plot means analysis.

```
                  CLASS LEVEL INFORMATION

CLASS      LEVELS  VALUES

BZ            2    CB SP
PLT           4    7 8 9 10
REP           8    1 2 3 4 5 6 7 8
SOURCE        2    CB SP
FEM          37    2 10 20 21 34 38 47 49 57 69 88 90 95 97
                   104 106 108 109 113 118 121 409 427 433
                   438 441 442 461 473 479 480 483 487 489
                   490 491 495
       Number of Observations in Data Set = 1917

DEPENDENT VARIABLE: HT8

SOURCE        TYPE I EXPECTED MEAN SQUARES
BZ            VAR(ERROR) + 39.9400 VAR(REP(BZ*PLT))
              + 319.5016 VAR(PLT(BZ))
              + Q(BZ,SOURCE,BZ*SOURCE,FEM(SOURCE),
              BZ*FEM(SOURCE),PLT*SOURCE(BZ),
              PLT*FEM(BZ*SOURCE))

SOR           VAR(ERROR) + 0.0025 VAR(REP(BZ*PLT))
               + 0.0017 VAR(PLT(BZ)) + Q(SOURCE,BZ*SOURCE,
               FEM(SOURCE),BZ*FEM(SOURCE),PLT*SOURCE(BZ),
               PLT*FEM(BZ*SOURCE))
```

Cont'd next page

Box 6.8 cont'd

```
BZ*SOR       VAR(ERROR) + 0.0025 VAR(REP(BZ*PLT))
             + 0.0017 VAR(PLT(BZ)) +
               Q(BZ*SOURCE,FEM(SOURCE),
             BZ*FEM(SOURCE),PLT*SOURCE(BZ),
             PLT*FEM(BZ*SOURCE))

FEM(SOR)     VAR(ERROR) + 0.0028 VAR(REP(BZ*PLT))
             + 0.0015 VAR(PLT(BZ)) + Q(FEM(SOURCE)
             ,BZ*FEM(SOURCE),PLT*SOURCE(BZ)
             ,PLT*FEM(BZ*SOURCE))

BZ*FEM(SOR)  VAR(ERROR) + 0.0028 VAR(REP(BZ*PLT))
             + 0.0015 VAR(PLT(BZ)) + Q(BZ*FEM(SOURCE)
             ,PLT*SOURCE(BZ),PLT*FEM(BZ*SOURCE))

PLT(BZ)      VAR(ERROR) + 39.9363 VAR(REP(BZ*PLT))
             + 319.471265 VAR(PLT(BZ))
             + Q(PLT*SOURCE(BZ),PLT*FEM(BZ*SOURCE))

REP(BZ*PLT)  VAR(ERROR) + 39.9327 VAR(REP(BZ*PLT))
             + Q(PLT*SOURCE(BZ),PLT*FEM(BZ*SOURCE))

PLT*SOR(BZ)  VAR(ERROR) + Q(PLT*SOURCE(BZ),
                PLT*FEM(BZ*SOURCE))

PLT*
FEM(BZ*SOR)  VAR(ERROR) + Q(PLT*FEM(BZ*SOURCE))
```

Box 6.9. Output from PROC GLM analysis of wide test study.

DEPENDENT VARIABLE: HT8

SOURCE	DF	SUM OF SQUARES	MEAN SQUARE	F VALUE	PR > F
MODEL	263	1569.3375	5.9670	23.73	0.0
ERROR	1653	415.5744	0.2514		
TOT CSS	1916	1984.9120			

R-SQUARE	C.V.	ROOT MSE	HT8 MEAN
0.660377	13.2458	0.50140426	3.78538341

SOURCE	DF	TYPE I SS	F VALUE	PR > F
BZ	1	820.5029	2063.06	0.0
SOURCE	1	0.9010	3.58	0.0585
BZ*SOURCE	1	0.2182	0.87	0.3516
FEM(SOURCE)	35	80.8742	9.19	0.0
BZ*FEM(SOURCE)	35	14.6511	1.67	0.0089
PLT(BZ)	4	357.6703	355.67	0.0
REP(BZ*PLT)	42	258.4202	24.47	0.0
PLT*SOURCE(BZ)	4	3.6264	3.61	0.0062
PLT*FEM(BZ*SOURCE)	140	32.4726	0.92	0.7272

SOURCE	DF	TYPE III SS	F VALUE	PR > F
BZ	1	790.8631	3145.76	0.0
SOURCE	1	0.1466	0.59	0.4410
BZ*SOURCE	1	0.2410	0.95	0.3305
FEM(SOURCE)	35	81.0920	9.21	0.0
BZ*FEM(SOURCE)	35	14.3921	1.63	0.0119

Box 6.10 Wide test PROC VARCOMP plot mean analysis.

EXPECTED MEAN SQUARES

Source	DF	Mean Sq.	σ^2_e	$\sigma^2_{f(s)}$	$\sigma^2_{ls(z)}$	$\sigma^2_{b(zl)}$	$\sigma^2_{l(z)}$	F.E.	F	F Formulae
BZ	1	820.398 (1)	1	9.2	159.8	39.94	319.5	$\Sigma\beta^2$	9.2	MS1/MS6
SOURCE	1	0.903 (2)	1	9.2	159.8	0.01	0.002	$\Sigma\Gamma^2$	1.0	MS2/MS8
BZ X SOURCE	1	0.222 (3)	1	9.2	159.8	0.01	0.002	$\Sigma(\beta\Gamma)^2$	0.2	MS3/MS8
FEM(SOURCE)	35	2.307 (4)	1	8.6	0.002	0.01	0.002	$\Sigma\phi^2$	5.5	MS4/MS5
BZ X FEM(S)	35	0.419 (5)	1	8.6	0.002	0.01	0.002	$\Sigma(\beta\phi)^2$	1.8	MS5/MS9
PLT(BZ)	4	89.353 (6)	1	9.2	159.8	39.94	319.5		12.7	(MS6+MS10)/ (MS7+MS8)
REP(BZ X PLT)	42	6.150 (7)	1	0.0	0.001	39.93			24.5	MS7/MS10
PLT X S(BZ)	4	0.921 (8)	1	9.2	159.7				4.0	MS8/MS9
PLT X FEM(BZ X S)	140	0.232 (9)	1	8.6	0.002				0.9	MS9/MS10
ERROR	1653	0.251 (10)	1							

Cont'd next page

Box 6.10 Cont'd

PROC GLM MEAN SQUARES

		TYPE I	F	TYPE III	F	TYPE IV	F
BZ	1	820.503	9.2	790.863	9.2	791.812	9.1
SOURCE	1	0.901	1.0	0.149	0.2	0.149	0.2
BZ X SOURCE	1	0.218	0.2	0.238	0.2	0.238	0.2
FEM(SOURCE)	35	2.311	5.5	2.316	5.7	2.316	5.7
BZ X FEM(S)	35	0.419	1.8	0.409	1.8	0.409	1.8
PLT(BZ)	4	89.418	12.7	85.957	12.2	86.890	12.3
REP(BZ X PLT)	42	6.153	24.5	6.157	24.5	6.156	24.5
PLT X S(BZ)	4	0.907	3.9	0.940	4.0	0.940	4.0
PLT X FEM(BZ X S)	140	0.232	0.9	0.232	0.9	0.232	0.9
ERROR	1653	0.251		0.251		0.251	

σ^2_e = Residual error variance; $\sigma^2_{f(s)}$ = Family within source variance; $\sigma^2_{ls(z)}$ = Location x source variance; $\sigma^2_{b(zl)}$ = Replication variance; $\sigma^2_{l(z)}$ = Planting location variance.

$\Sigma\beta^2$, $\Sigma\Gamma^2$, $\Sigma(\beta\Gamma)^2$, $\Sigma\phi^2$, $\Sigma(\beta\phi)^2$, are fixed effects for breeding zone, source, breeding zone x source, family, and breeding zone x family, respectively.

Box 6.11. Wide test specified hypothesis tests using GLM.

```
DEPENDENT VARIABLE: HT8
Ho: TYPE I MS USING PLT(BZ) AS ERROR TERM

SOURCE      DF       TYPE I SS     F VALUE     PR > F
BZ           1    820.50298921      9.18       0.0388

Ho: TYPE I MS USING PLT*SOURCE(BZ) AS ERROR TERM

SOURCE      DF       TYPE I SS     F VALUE     PR > F
SOURCE       1    0.90105007        0.93       0.6181

Ho: TYPE I MS USING PLT*SOURCE(BZ) AS ERROR TERM

SOURCE      DF       TYPE I SS     F VALUE     PR > F
BZ*SOURCE    1    0.21828108        0.24       0.6512

Ho: TYPE I MS USING BZ*FEM(SOURCE) AS ERROR TERM

SOURCE         DF    TYPE I SS     F VALUE     PR > F
FEM(SOURCE)    35   80.87425484     5.52       0.0001

Ho: TYPE I MS USING PLT*FEM(BZ*SOURCE) AS ERROR TERM

SOURCE             DF   TYPE I SS   F VALUE   PR > F
BZ*FEM(SOURCE)     35  14.65114528   1.81     0.0082

Ho: TYPE I MS FOR PLT*FEM(BZ*SOURCE) AS AN ERROR TERM
SOURCE
```

6.2.3 Multiple comparisons

In many forest genetics experiments, statistical tests of specific differences among genetic entries are of interest. Multiple comparison methods may be used to develop these tests. There is a great variety of multiple comparison methods, and a large body of literature exists on their advantages and disadvantages.

The major concern, in applying multiple comparison methods, lies in controlling the TYPE I (experimentwise) error rate; that is, the probability of declaring a pair of means different when in fact they are equal. The most commonly used methods, LSD and Duncan's multiple range test, are frequently criticized for inflating the TYPE I error rate. When applicable, the use of *a priori* planned comparisons are preferred, and should be used when specific hypotheses are suggested by the treatment structure.

In Box 6.12, I have presented results from a six-year height PROC GLM analyses of a Weyerhaeuser study established with 16 full-sib families in 6 replications of 4 trees on 16 planting locations. The results of the Duncan's multiple range test are in Box 6.13.

To show the development of planned comparisons, and the use of CONTRAST within PROC GLM, I developed the SAS® stack at the top of Box 6.14. The trick here is to make sure that the order of the treatment contrasts (FEM in this case) agree with the order of the specified value constants below the CONTRAST statement (6 6 6 6 -2 ...). In this example, I constructed the contrast of the top 4 families,

Box 6.12. Run stack for GLM with multiple comparisons.

```
LIBNAME PLOTM 'I:\';
TITLE 'CB SUPPLEMENTAL SITES MEAN
ANALYSIS MULTIPLE COMPARISONS';
PROC GLM DATA=PLOTM.CB88;
CLASS PLT REP FEM;
MODEL HT6=PLT REP(PLT) FEM FEM*PLT;
MEANS FEM / DUNCAN;
RUN;
```

CB SUPPLEMENTAL SITES MEAN ANALYSIS MULTIPLE COMPARISONS.

GENERAL LINEAR MODELS PROCEDURE

CLASS LEVEL INFORMATION

CLASS	LEVELS	VALUES
FEM	16	1 9 16 20 31 47 48 53 60 61 68 75 76 88 98 102
PLT	16	25 26 27 28 29 30 31 32 33 34 35 36 37 38 39 40
REP	6	1 2 3 4 5 6

NUMBER OF OBSERVATIONS IN DATA SET = 1536

DEPENDENT VARIABLE: HT6

SOURCE	DF	SUM OF SQUARES	MEAN SQUARE	F	PR > F
MODEL	335	1207.5421	3.6045	40.09	0.0
ERROR	1200	107.8871	0.0899		
CORRECTED TOTAL	1535	1315.4292			

R-SQUARE	C.V.	ROOT MSE	HT6 MEAN
0.9179	12.1005	0.2998	2.4779

Cont'd next page

Box 6.12 cont'd

SOURCE	DF	TYPE I SS	F VALUE	PR > F
PLT	15	1065.2535	789.90	0.0
REP(PLT)	80	71.0870	9.88	0.0
FEM	15	40.3849	29.95	0.0
FEM*PLT	225	30.8166	1.52	0.0001

SOURCE	DF	TYPE III SS	F VALUE	PR > F
PLT	15	1065.2535	789.90	0.0
REP(PLT)	80	71.0870	9.88	0.0
FEM	15	40.3849	29.95	0.0
FEM*PLT	225	30.8166	1.52	0.0001

determined from results in a different field test, with the rest of the families in the test analyzed here.

It should be emphasized, that while planned comparisons are the preferred method for testing specific differences, the comparisons should be *a priori* comparisons. If this is not the case, one could be justifiably accused of "shopping" among the treatments for differences.If planned comparisons are not possible, a multiple comparison method should be used.

6.2.4 Computing genetic correlations

Relationships among several traits are often of interest in forest genetics studies, particularly those designed to estimate components of variance. Estimates of genetic and environmental covariances are required for development of multiple trait selection indexes, and for

Box 6.13. Duncan's multiple range test for 16 family study.

```
DUNCAN'S MULTIPLE RANGE TEST FOR VARIABLE: HT6

NOTE: This test controls the Type I comparisonwise error
rate, not the experimentwise error rate.

                      ALPHA=0.05  DF=1200  MSE=.0899059
NUMBER OF MEANS         2         3         4         5         6
CRITICAL RANGE 0.085974 0.090407 0.093260 0.095403 0.097206

NUMBER OF MEANS         7         8         9        10        11
CRITICAL RANGE 0.098680 0.099886 0.100883 0.101714 0.102416

NUMBER OF MEANS        12        13        14        15        16
CRITICAL RANGE 0.103014 0.103529 0.103975 0.104366 0.104709

MEANS WITH THE SAME LETTER ARE NOT SIGNIFICANTLY DIFFERENT.

DUNCAN GROUPING          MEAN       N  FEM
         A              2.75010    96  31
         A
B        A              2.73104    96  76
B        A
B        A              2.66552    96  47
B
B                       2.64406    96  1
B
B                       2.64010    96  20
B
B                       2.63594    96  9

         C              2.43208    96  53
         C
D        C              2.39031    96  98
D        C
```

Cont'd next page

Box 6.13 cont'd

DUNCAN	GROUPING	MEAN	N	FEM
D	C	2.38417	96	48
D	C			
D	C	2.38115	96	102
D	C			
D	C	2.37750	96	60
D	C			
D	C	2.37552	96	88
D	C			
D	C	2.33990	96	61
D	C			
D	C	2.33896	96	68
D				
D	E	2.32240	96	16
	E			
	E	2.23833	96	75

determining the influence that selection for one trait may have on another.

Multivariate analysis of covariance provides one means (albeit cumbersome) for estimating covariances among traits. A second method uses the algebraic relationship of the variance of a sum of two variables. The individual traits and their sum(s) use univariate analyses, and the covariances are calculated. I have included an example in Box 6.15 that compares results using multivariate and univariate analyses. Note that the estimates of genetic correlations (r_g)are the same. The univariate approach is much less consumptive of computer time and resources.

Box 6.14 Using planned comparisons (contrasts).

```
* USING PROC GLM FOR CONTRASTS;
LIBNAME PLOTM 'I:\;
DATA FBRP25;
SET PLOTM.CB88;
WHERE PLT EQ 25;
IF FEM EQ 1 THEN FEM=155;
IF FEM EQ 9 THEN FEM=1;
IF FEM EQ 31 THEN FEM=2;
IF FEM EQ 47 THEN FEM=3;
IF FEM EQ 76 THEN FEM=4;
PROC GLM ORDER = FORMATTED;
CLASS FEM REP;
MODEL MX2 = REP FEM;
CONTRAST 'TOP NI GT VS REST ' FEM
6 6 6 6 -2 -2 -2 -2 -2 -2 -2 -2 -2 -2 -2 -2 -2;
TITLE 'CB SUP SITES PLT 25 ONLY GLM WITH CONTRAST';
RUN;

CB SUP SITES PLT 25 ONLY GLM WITH CONTRAST

Class Level Information
Class     Levels     Values
FEM        16        1 2 3 4 16 20 48 53 60 61 68 75 88 98
                     102 155
REP         6        1 2 3 4 5 6
    Number of observations in data set = 96

        CB SUP SITES PLT 25 ONLY GLM WITH CONTRAST
            General Linear Models Procedure

Dependent Variable: MX 2   HT6 MEAN m
```

Cont'd next page

Box 6.14 cont'd

Source	DF	Sum of Squares	Mean Square	F Value	Pr>F
Model	20	15.26063	0.763031	8.53	0.0001
Error	75	6.70838	0.089445		
Corrected Total	95	21.96902			

R-Square	C.V.	Root MSE	MX2 Mean
0.694643	13.64433	0.299074	2.19192708

Source	DF	Type I SS	Mean Square	F Value	Pr>F
REP	5	12.28481554	2.45696311	27.47	0.0001
FEM	15	2.97582104	0.19838807	2.22	0.0126

Source	DF	Type III SS	Mean Square	F Value	Pr>F
REP	5	12.28481554	2.45696311	27.47	0.0001
FEM	15	2.97582104	0.19838807	2.22	0.0126

Dependent Variable: MX2 HT6 MEAN m

Contrast	DF	Contrast SS	Mean Square	F Value	Pr>F
TOP IN GT VS REST	1	1.74611328	1.74611328	19.52	0.0001

6.2.5 More on missing cells and unbalanced data

It is difficult to generalize as to the appropriate analytical method for dealing with data sets containing missing cells. SAS®, for example, provides nine methods for carrying out analysis of variance. Since degree of unbalance may vary markedly, results and consumption of computing resources can be significantly affected by choice of method.

Box 6.15. Two methods of computing genetic correlations.

```
* PROC ANOVA WITH MANOVA FOR MULTIPLE TRAIT ANALYSES;
LIBNAME PLOTM 'I:\';
DATA FBR;
SET PLOTM.CB88;
PROC ANOVA;
CLASS FEM PLT REP;
MODEL MX1 MX2=PLT REP(PLT) FEM PLT*FEM;
MANOVA H=FEM PLT*FEM / PRINTH PRINTE;
TITLE 'CB SUP SITES MEAN ANALYSIS PROC ANOVA & MANOVA';
RUN;

* PROC VARCOMP USING SUM OF TRAITS TO OBTAIN
CROSSPRODUCTS;
LIBNAME PLOTM 'I:\';
DATA FBR;
SET PLOTM.CB88;
* SUM TRAITS;

HT4HT6=MX1 + MX2;
PROC VARCOMP METHOD=TYPE1;
CLASS FEM PLT REP;
MODEL MX1 MX2 HT4HT6=PLT REP(PLT) FEM PLT*FEM
TITLE 'CB SUP SITES MEAN ANALYSIS PROC VARCOMP SUM
     VARIABLES FOR CP';
RUN;
```

CALCULATION OF GENETIC CORRELATIONS ANOVA WITH
MANOVA VS VARCOMP

	HT4 MS	HT6 MS	HT46 MCP	Expected Mean Squares & Cross Products
FEM	0.5829	2.6935	1.2153	$\sigma^2e + 6\ \sigma^2fp + 96\ \sigma^2f$
PLT*FEM	0.0401	0.1371	0.0643	$\sigma^2e + 6\ \sigma^2fp$
ERROR	0.0276	0.0898	0.0429	σ^2e
	σ^2f H4	σ^2f H6	σf H4H6	r_g
ANOVA	0.005655	0.026629	0.01199	0.98

Cont'd next page

Box 6.15 cont'd

```
                                         σ²f(H4+H6)
VARCOMP              0.005655 0.026629    0.05626       0.98
                                σf H4H6   0.01199

VAR(x+y)=VARx + VARy + 2COVxy
COVxy=(VAR(x+y) - VARx - VARy)/2
```

Box 6.16 shows the results from analyses of the same data set used in Box 6.12. To compare several methods, the data set was first analyzed using PROC ANOVA (which should only be used for balanced data), and then processed, after eliminating 14 cells that had only 1 tree, using PROC GLM Type IV SS, and PROC VARCOMP Methods TYPE1, MIVQUEO, ML, and REML. See page 968 of SAS® Institute Inc. SAS®/STAT Guide for Personal Computers Version 6 Edition, for a description of these methods.

The F tests and variance component estimates in Box 6.16 are very similar, and any of the methods used would lead to the same statistical conclusions. This is true even for the F tests for the unbalanced data set using PROC ANOVA. It should be emphasized, however, that this specific example had few missing cells in relation to the total degrees of freedom available.

Again, it is difficult to make general recommendations about analytical methods used for data sets with varying degrees of unbalance. It has been my experience that low levels of unbalance can be adequately dealt with using VARCOMP TYPE I. Moderate or severe levels of unbalance should more properly employ GLM Type IV SS.

Box 6.16. Comparisons of SAS® procedures for analyses of balanced and unbalanced data sets.

BALANCED PROC ANOVA

SOURCE	df	ANOVA MEAN SQ	ANOVA F VALUE
PLT	15	71.000	69.4
REP(PLT)	80	0.888	10.0
FEM	15	2.694	19.7
PLT*FEM	225	0.137	1.5
ERROR	200	0.089	

14 MISSING CELLS (REMOVED MANUALLY FROM DATA SET FOR CELLS WITH ONLY 1 TREE PER FAMILY-REP.)

		ANOVA		VC TYPE I		GLM IV	
SOURCE	df	MS	F	MS	F	MS	F
PLT	15	70.514	69.7	70.514	69.7	70.319	69.9
REP(PLT)	80	0.873	10.1	0.873	10.1	0.866	10.0
FEM	15	2.644	18.9	2.579	18.3	2.528	17.9
PLT*FEM	225	0.140	1.62	0.141	1.62	0.141	1.6
ERROR	1186	0.087		0.087		0.087	

VARIANCE COMPONENT ESTIMATES FROM RUNS USING PROC VARCOMP.

	TYPE I	MIVQUE	ML	REML
$\sigma^2 f$	0.0256	0.0261	0.0255	0.0256
$\sigma^2 fp$	0.0091	0.0096	0.0091	0.0091

Literature Cited

Bartlett, M. S. 1947. The use of transformations. *Biometrics* 3(1):39-52.

Cochran, W. C. 1947. Some consequences when the assumptions for the analysis of variance are not satisfied. *Biometrics* 3(1):22-38.

Eisenhart, C. 1947. The assumptions underlying the analysis of Variance. *Biometrics* 3(1):1-21.

Freund, R. J., R. C. Littell, and P. C. Spector. 1986. *SAS® System for Linear Models*. SAS®Institute Inc., Cary, NC.

Kennedy, B. W. 1981. Variance component estimation and prediction of breeding values. *Can. Jour. of Genetics and Cytology* XXIII(4):565-578.

SAS® Institute Inc. *SAS®/STAT Guide for Personal Computers Version 6 Edit*. SAS® Institute Inc., Cary, NC, 1987. 1028 p.

Schultz, E. F., Jr. 1955. Rules of thumb for determining expectations of mean squares in analysis of variance. *Biometrics* 11:123-135.

Searle, S. R. 1971. Topics in variance component estimation. *Biometrics* 27:1-76.

Snedecor, George W., and W. G. Cochran. 1976. *Statistical Methods*, Sixth Edition. Iowa State Univ. Press. Ames, Iowa. 593 p.

Chapter 7
Estimating Yield: Beyond Breeding Values

G.S. Foster[1]

Tree improvement, as a subdivision of genetics, has several unique features when compared with genetic improvement programs for other plant and animal species.

7.1 Introduction

Historically, breeding plans have been developed for domesticated plant and animal species with relatively short generation times and few offspring per parent. Conversely, most forest tree species have relatively long generation times (10 to 20 years) and have an enormous individual capacity to produce offspring (i.e., within the productive lifetime of a seed orchard). The economic lifespan of a stand of forest trees ranges from 25 to more than 100 years, while the economic lifespan for most domesticated crop plants is rarely more than two years and little more for most domesticated animals. In addition, forest managers plan on harvesting less than half of the individuals initially established in a stand of trees; therefore, inter-genotypic competition assumes major importance in shaping individual

[1]*Research Geneticist, USDA Forest Service, Southern Forest Experiment Station, Alabama A&M University, Normal, Alabama*

L. Fins et al. (eds.), Handbook of Quantitative Forest Genetics, 229–269.

phenotypes and inducing mortality. These features, as well as others that are unique to forest trees, present a need to alter genetic gain prediction systems that were originally developed for insects, domesticated animals, or agronomic crops (e.g., Falconer 1981).

7.2 Genetic Tests

Genetic test design in forestry developed from theoretical and empirical research with insects, domesticated animals, and agronomic crops. Tree breeders have utilized numerous field-test designs over the years, ranging from large plots of up to 100 trees per provenance or family to single tree plots. Several research studies have been conducted that compare test designs for efficiency of family mean estimation (e.g., Lambeth et al. 1983; Loo-Dinkins and Tauer 1987; Wright and Freeland 1960). Results from these studies seem generally to favor very small plot sizes (e.g., one to five trees) with some advantage to non-contiguous plots compared with row plots.

However, a major problem with these results is that the most important trait for improvement is often wood yield per unit area at rotation age. Much effort has been expended by forest geneticists to find a trait or traits that can be measured at a relatively early age and will serve as an efficient indirect selection criterion. Based on forest mensurational results, the most efficient predictors of current or future stand growth and yield may not be the mean of all living trees but rather the mean of a subset of living trees *after* competition has started (e.g., dominant-codominant height versus mean of all living trees to estimate site quality) (Clutter et al. 1983). Competition among trees within a stand arises from the allocation of limited resources (e.g., light, moisture, nutrients) to growing trees. As inter-tree competition intensifies, frequency distributions, means, and variances may be affected; hence, genetic gains estimated from results using small-plot progeny tests may be biased.

7.2.1 Competition effects on frequency distributions

The basic assumption of a normal distribution of phenotypic and genotypic values as a basis for selection and genetic gain calculation (Bulmer 1985) is questionable in populations of forest trees. The underlying distribution of height, diameter at breast height (d.b.h.), and volume for a forest stand changes with stand age due to inter-tree competition and mortality (Cannell 1984; Ford 1975; Mohler et al. 1978). Seedling size at outplanting and for the first few years in the field approximates a normal distribution; however, as inter-tree competition intensifies, the distributions become more and more skewed to the lower end (Mohler et al. 1978; Cannell 1984). Finally, competition becomes so intense that self thinning starts. Eventually enough of the smaller, more suppressed trees die so that the distribution may once again approximate a normal distribution (Mohler et al. 1978) or even a slightly bimodal distribution (Cannell 1984; Ford 1975).

Normality is also a function of sample size, and large sample theory indicates a tendency toward normality in many large populations. However, non-normal distributions may be common in forestry when genetic gain is estimated using small numbers of families and selection is based on family-means.

Genetic gain may be estimated as follows (Falconer 1981):

$$\mathbf{G = S\ h^2} \qquad \text{Eq. 7.1}$$

where,

- G = genetic gain
- S = selection differential
- h^2 = heritability.

Additionally, with a normal distribution,

$$\mathbf{S = i\ \sigma_P}$$

where,

- i = intensity of selection[2] = z/p
- z = height of the ordinate at the point of truncation of a normal distribution
- p = proportion selected
- σ_P = phenotypic standard deviation

In cases where the distribution is not normal, the value of S should be calculated from the data rather than estimated from intensity of selection (i) and phenotypic standard deviation (σ_P).

---

[2]Editor's Note: Lerner (1958) in "The Genetic Basis of Selection" defined selection intensity as "the percentage of the population permitted to reproduce itself." The definition given here is equivalent to Lerner's standardized selection differential ($\bar{\imath}$), which Falconer (1981) called "the intensity of selection." This discrepancy in terminology can lead to confusion, and suggests that, as Foster has, the author clearly indicate which definition is being used.

7.2.2 Plot size

The size of a plot (representing a provenance, family, or clone in a particular block) depends on the goal of the test. As stated above, small plot sizes (e.g., one to five trees) are favored when the goal of testing is the estimation of family means of all living trees. Since the goal of tree improvement is often to enhance volume yield per unit area, the plot size should reflect stand dynamics. Once competition initiates the process of crown class development, small plots are no longer an accurate representation of a forest stand, because the size distributions of trees within the plots may be totally abnormal. For instance, all trees in the family plot could be either all dominants or all suppressed.

Mensurationists have found that the average height of the dominant-codominant trees is one of the variables needed to accurately predict stand volume as well as future growth (Clutter et al. 1983). The diameter distribution can be predicted from the dominant-codominant height, stand age, and number of trees per unit area. The number of sample trees required to estimate dominant-codominant height can vary, but the 100 tallest trees/acre is a common number of sample trees (Buford and Burkhart 1985, 1987). The average number of trees per plot that would be needed to include the tallest 100 trees/acre depends on the initial spacing and the number of trees per plot (Table 7.1).

For example, an 8 x 8 ft initial spacing would yield 680 trees/acre; therefore, the tallest 100 would represent 14.7% of the original trees. In order to average one dominant-codominant tree per plot, at least seven trees would have to be included in each plot (Table 7.1). With 50-tree plots, 7 dominant-codominant trees on the average would be included and could be used to calculate the dominant-codominant height of the forest stand.

The tree breeder is faced with a dilemma regarding plot size and testing objectives. To date, the best solution seems to include the establishment of a series of tests. For screening large numbers of families or clones, an initial group of tests is installed using small plots, and the trees are measured prior to crown closure. Then new trials using large plots are initiated for testing long-term yield potential of only the better families or clones (Libby 1987).

Table 7.1. Number of dominant-codominant (DC) trees in a plot as a function of plot size, assuming that the tallest 100 trees per acre are DC and with an 8x8 ft. spacing.

Trees/plot	DC Trees/plot
4	0.6
5	0.7
6	0.9
7	1.1
8	1.2
9	1.3
10	1.5
25	3.7
50	7.3
100	14.7

7.2.3 Plot configuration

A provenance, family, or clone plot can vary not only in size, but also in configuration. Some tree breeders, especially in the northwestern United States, Canada, and Europe, utilize single tree plots. In this case, a single tree per provenance, family, or clone is placed within each replication. Typically, small plots (e.g., 1 < size ≤ 10 trees) are placed in rows, while larger plots (e.g., size > 10 trees) are configured in square or rectangular plots. In these designs, the trees within a plot are generally contiguous (i.e., adjacent to each other), although non-contiguous multiple tree plots have been used more frequently since the late 1970's (Lambeth et al. 1983; Libby and Cockerham 1980).

In the free-to-grow status of the early years of a genetic test, inter-tree competition does not affect growth of the test trees; however, once inter-tree competition begins, means and variances may be biased. Inter-family and intra-family (provenance or clone) competition varies tremendously with changes in plot configuration. The desired level of inter-genotypic competition is dictated by the objectives of the test. For example, if families or clones will be mixed for reforestation, some type of single tree plot or non-contiguous multiple tree plot should be favored for the genetic tests, because this simulates future stands. Conversely, if the families will be planted in pure (single) family blocks for reforestation, then the genetic tests should utilize larger pure plots to simulate a pure stand.

Research results published to date seem to indicate that plot configuration exerts only a small effect on mean estimation (hence family ranking), yet large changes are detected in variances. Foster (1989) summarized published results and concluded that average family performance was relatively stable across a range of inter-family competition. He also noted that the variance among families increased dramatically when changing the plot configuration from: a large block plot to a row plot, a large block plot to a non-contiguous plot, or from a row plot to a non-contiguous plot. Therefore, family ranking for selection should be relatively stable among different plot configurations, providing for efficient selection; yet genetic gain estimates would be inflated in a small plot experiment due to the confounding of genetic variance and competitional variance.

7.2.4 Deployment strategy

As noted above, there must be a relationship between the testing strategy and the deployment strategy for reforestation with the improved material. The variation among pure family blocks in both progeny tests and reforestation reflects only genetic variance with no confounding of inter-family competitional variance. Yet the variation among family row plots, non-contiguous multiple tree plots, or single tree plots confounds genetic variation and inter-family competitional variation.

Theoretically, the competitional variation should be part of the realized genetic gain since mixed family stands include this type of variation in genetic tests and commercial stands. However, the tests include families that are rogued from the seed orchard prior to reforestation. The reduction in families to include only the superior ones in the resultant stands will cause some reduction in the genetic variance. It may also cause a change in the competitional variance, since the inferior families that would comprise a large part of the suppressed crown class in the test will not be represented in the commercial stands.

Superior tree or family selection must include an assessment of the role of competition and its interaction with the genotype(s) involved. Area Potentially Available (APA) is one promising index of available growing space or individual-tree competitive status (Nance et al. 1983). APA provides an estimate of the available growing space, and the relationship between APA and individual tree size (averages of APA and tree sizes for family status) provides an assessment of how well a particular genotype or family utilizes the area that it has. Theoretically, an assemblage of genotypes that efficiently use their allocated growing space and have a relatively large size (crop ideotype) would produce the largest yield per acre (Cannell 1982). Little information is yet available for older studies to fully assess this procedure.

Tree breeders desperately need research results to help improve genetic test designs so that the results can be translated into realized genetic gain estimates.

7.3 Part-Record Selection

The use of part-records is one aid to selection that seems to be applied universally in tree improvement. Part-record selection refers to the practice of selection based on something less than total production (Turner and Young 1969). Total production for yield in forestry refers to production at rotation age. Virtually no tree species has a rotation sufficiently short enough for tree breeders to wait for rotation-length test results before selection. Tree breeders, therefore, generally choose some age less than rotation length to make their final selections, hoping that genetic gain in rotation-length production will be forthcoming.

Part-record selection is often termed indirect selection for which the theoretical expectations are given by Falconer (1981) and Turner and Young (1969).

$$CR_Y = i\, h_X\, h_Y\, r_G\, \sigma_{PY} \qquad \text{Eq. 7.2}$$

where,

CR_Y	=	correlated response in trait Y (full production trait)
i	=	intensity of selection
h_X	=	square root of heritability for trait X (part-record trait)
h_Y	=	square root of heritability for trait Y
r_G	=	genetic correlation between traits X and Y
σ_{PY}	=	phenotypic standard deviation for trait Y

Examples of part-record selection are prevalent in the dairy industry (e.g., Van Vleck and Henderson 1961), egg industry (e.g., Van Vleck and Doolittle 1964), and chicken broiler industry (Martin et al. 1953). Forest geneticists have also used part-record selection (or indirect selection) to predict the correlated response at a later age (Foster 1986; Nanson 1970; Squillace and Gansel 1974). Unfortunately, only a limited number of genetic tests of forest trees have reached rotation age; therefore, only a few estimates of genetic parameters at rotation age are even available. Genetic parameter estimates are needed that relate juvenile (part-record) traits in small plot tests with rotation age traits in large plot tests. Without accurate estimates of genetic parameters and realized genetic gain with the same population, the utility of part-record selection in tree improvement remains untested.

7.4 Deployment Strategy

7.4.1 Seed orchard option

Most tree improvement programs utilize seed from seed orchards to deploy their production population. Several factors governing seed production and seedling deployment pose strong effects (both cultural and genetic) on realized genetic gain from commercial stands.

7.4.1.1 Number and fecundity of clones in orchard

The number and relative fecundity (i.e., ability to produce pollen and seeds) of parental clones in a seed orchard can have a large impact on realized genetic gain from improved stands arising from seed orchard seed. If few parental clones are included in the orchard, then the simplifying assumption of random mating (panmixia) may be violated due to non-random pollen dispersal (since pollen flow is distance dependent), non-uniform dispersal of ramets of each parental clone, inequality of pollen and female flower production among parental clones, and unequal numbers of ramets per parental clone. Violation of the assumption of random mating leads to necessary changes in the genetic gain calculation.

The technique of adjusting expected breeding values, as shown in Box 7.1, can be used if the seed from only a proportion of the parental clones in the orchard is harvested. Harvesting the seed from only the

Box 7.1 Adjustment of expected breeding value for varying proportions of parental clones in seed orchard.

One method to account for either varying proportions of ramets of each parental clone in a seed orchard or varying fecundity is to weight the breeding value of each parental clone by the proportion of ramets or relative fecundity. For example:

Parental Clone	Proportion of ramets	Breeding Value	Weighted breeding value
A	0.10	100	10
B	0.50	110	55
C	0.20	105	21
D	0.20	115	23
Total	1.00		109

Therefore, the breeding value of the offspring population would be 109 instead of 107.5 as if each parental clone had equal representation.

best clones within a seed orchard has become a common practice as orchards reach full production and seed production exceeds demand. Additionally, some seed orchard managers will deploy the seed from different subsets of the parental clones in an orchard to different sites depending, for example, on site index or disease hazard rating. Finally, the technique of weighting the breeding value can be used following the roguing of a seed orchard, since the seed from the rogued parental clones has been deleted from the harvest. The pollen contribution from the rogued parental clones will still be evident for up to one or two (for pines) harvests following the roguing.

For genetic gain calculations, heritability estimates obtained from test results may have to be adjusted if there are too few clones in the orchard even if random mating is occurring. The genetic expectations depend on the degree of relationship among the trees (Hallauer and Miranda 1981).

Variance among families = covariance among sibs

Var. of half-sib families = $Cov_{HS} = \frac{1 + F}{4} V_A + \ldots$

Var. of full-sib families = $Cov_{FS} = \frac{2 + F}{4} V_A + \ldots$

(with the assumption of no dominance or epistasis)

where,

F = inbreeding coefficient

V_A = additive genetic variance

The coefficients of the additive genetic variance then depend on the effective number of males in the population and the amount of inbreeding among parents. Following Squillace (1974), the coefficients of the additive genetic variance (as a function of number of males and F) can be estimated as:

Selfing (%)	Number of equally effective males			
	5	10	15	20
10	0.34	0.32	0.31	0.30
0	0.30	0.27	0.26	0.26

If the number of parents in a seed orchard is small (and particularly if natural selfing occurs), then the seed gathered from each parental clone will have an average degree of relationship that is intermediate between a full-sib family and a half-sib family. Heritabilities, that are used to calculate expected genetic gain, must be adjusted to reflect the expected degree of family relationship.

7.4.1.2 Pollen contamination in a seed orchard

An influx of pollen into the seed orchard from a surrounding forest stand (termed pollen contamination) can significantly reduce realized genetic gain from a seed orchard. Marker systems such as isozyme analysis have provided methods whereby pollen contamination levels in seed orchards can be checked (Adams and Birkes 1989). The degree of pollen contamination from such analyses has been surprisingly high, 21-91% across three reported species (Adams and Birkes 1989).

For growth traits, pollen contamination acts by compromising the superior genetic quality of the orchard seed; but the greatest threat comes from compromises in phenology of the seed crop. Changes in the breeding value of the seed for growth traits due to contaminating pollen can be handled in the same manner as described in Section 7.4.1.1 and Box 7.1. The average breeding value of the surrounding stand and frequency of resultant pollination in the orchard would have to be known. Alteration of the seedling phenology may result if a seed orchard from one breeding zone were located in another breeding zone. This result becomes especially problematic in mountainous areas or with large shifts in latitude. Seedlings arising from such a widely relocated seed orchard may lack the cold hardiness needed for the intended breeding zone, yet be too slow growing for the breeding zone where the orchard is located.

7.4.1.3 Relatedness in seed orchard

Seed yield, seed germination rate, and seedling growth rate may decline if parental clones in the seed orchard are closely related or if ramets of the same clone are too close and they inter-breed frequently. Related parental clones were not a problem in first generation seed orchards but may become more so with successive generations of seed orchards.

Inbreeding depression appears to be severe with most cross-breeding forest tree species. In a slash pine (*Pinus elliottii* Engelm.) study, Squillace and Kraus (1962) noted average decreases of 5 sound seeds per cone, 8% in germination rate, and 4.3% in 4-month seedling height per 0.1 increase in F. Similarly, Sorensen (1971) found that Douglas-fir [*Pseudotsuga menziesii* (Mirb.) Franco] sets about 10% as many seeds after self-pollination compared with cross-pollination. With loblolly pine (*Pinus taeda* L.), Sniezko and Zobel (1988) found a 1.0% and 1.2% decrease in seedling height and diameter, respectively, for each 0.1 increase in F.

Based on pollen dispersal studies, ramets of the same clone are spaced a minimum distance apart of 90 ft in southern pine seed orchards. This "90 Foot Rule" becomes increasingly more difficult to maintain as the number of clones in a seed orchard decreases or as related parental clones are included in the same seed orchard. Increasingly, seed orchard managers are dividing their seed orchards into blocks, with related parental clones being confined to different orchard blocks.

Loss of seed germination rate or seedling growth vigor will negatively impact a reforestation program by the production of proportionately more lower grade seedlings in the nursery. A seedling grading program may remove the obvious culls, but a general decline in vigor for a whole seedling crop might go unnoticed and would reduce growth and yield of the resultant stands. These negative effects will cause a deviation of realized genetic gain from expected genetic gain solely due to cultural practices.

7.4.1.4 Cone collection and seedling production strategies

An initial phase of the deployment strategy begins with cone collection in the seed orchard and continues with seedling production in the nursery. Cones can be collected by (1) bulk mixes of parental clones (i.e., all parental clones combined at random), (2) groupings of parental clones (i.e., two or more clones per group), or (3) pure parental clones (i.e., open-pollinated or control-pollinated families).

In the nursery production phase, seedlings can be grown in each of the same three groups. Each of the three methods has a profound impact on future deployment options. Potential genetic gain may be facilitated or hampered by deployment activities during seed collection and processing as well as seedling production.

The most common method of cone collection and seedling production is by bulk mixes. In this method, the cones are mixed at random during the harvesting operation, hence the seeds are mixed during processing. From a genetic standpoint, the resultant forest stands are more heterogeneous than natural stands, which may invoke a higher level of genetic homeostasis (the property of a population to equilibrate its genetic composition and to resist sudden changes) (Lerner 1954; Stern and Roche 1974). The frequencies of alleles derived from particular parents would be in proportion to the fecundity and number of ramets of the parental clones in the seed orchard. Seed size often varies by parental clone (Cannell et al. 1978); therefore, mixing cones may actually hamper efficient cone and seed processing due to a wider range in seed size. Furthermore, many nursery managers have their seed sized prior to sowing in the nurserybed to increase the efficiency of sowing and seedling production (Barnett and Dunlap 1982). Seed sizing may actually separate open-pollinated seed lots rather than leave them mixed (Silen and Osterhaus 1979). Seedling size is partially conditioned by genetic effects (Barnett and McLemore 1984); therefore, mixing seedlots may cause more variability in seedling size in the nursery and could potentially cause some families to be culled more than others.

With increasing frequency, orchard managers are collecting cones in groups of parental clones. These groups may originate from a variety of goals, including differential growth rates, disease resistances, and environmental sensitivities. As seed needs are exceeded by an organization, certain parental clones may even remain unharvested in some years. Groupings of similar parental clones provide much more flexibility for the forest manager to match genetic material with production needs. Genetic homeostasis may be reduced; yet the reduction need only be for the direct selection trait and other highly correlated traits. A wide range of variability should remain for uncorrelated traits. Unless parental clones are grouped by seed size or nursery growth, the lack of uniformity will cause the same production problems as noted above with bulk mixes.

Beginning in the late 1970's, a few forestry companies began to experiment with cone collection and nursery production with separate parental clones (i.e., open-pollinated families). The cone collection process is slowed somewhat; but nursery managers feel that the enhanced uniformity within a seedlot, in regard to speed of germination and response of seedlings to cultural manipulations, more than pays for the increased collection costs. The forest manager still maintains the option to either mix or maintain separate seedling lots during reforestation. With increasing frequency in the southeastern United States, nursery managers are requesting pure open-pollinated family seedlots.

7.4.1.5 Planting strategy

The same range of deployment strategies is being utilized for planting as for cone collection and seedling production. Organizational strategies range from bulk mixes to groupings or pure family plantings.

In the agronomic crop literature, numerous studies have compared pure plot and mixed plot performance of genetic entities (e.g., varieties, families, and clones). Trenbath (1974) summarized biomass yields from 344 mixtures presented in published results. He found that more than half (61%) of the binary mixtures equalled or exceeded the average yield of the pure plots (overcompensation), and 24% of the mixtures yielded more than the better yielding component. Only 13% of the mixtures yielded less than their lowest pure plot component.

Few studies of growth and yield of pure plots versus mixed plots have been published using forest tree species, and most of these studies utilized very young growth data. In his summary of published results, Foster (1989) found no consistent evidence for either superior or inferior average growth of pure plots versus mixed family plots. There were specific instances of significant interactions between certain families, however. Some large forestry companies in the southeastern United States have been allocating a significant portion of their annual reforestation effort to mosaics of pure family plots (e.g., Gladstone et al. 1987), in which a large acreage is planted to several open-pollinated families. Each family is confined to a large pure plot (usually $\leq$50 acres) within the total area. Reasons for this type of deployment strategy include: (1) enhanced understanding of the qualities of each family, (2) better matching of families to specific sites, and (3) more uniformity of raw material delivered to the mills. To date, no one knows the rotation-length result of this type of deployment versus mixed family stands. This issue requires immediate attention from researchers.

7.4.2 Clonal option

Clonal reforestation is being pursued with an increasing number of forest tree species worldwide. Many of the same deployment issues are important in clonal forestry as in seeding-related reforestation, but with a potentially greater impact on realized genetic gain.

7.4.2.1 Number of clones

The number of clones included in an improved population and deployed to a forest stand may have a major impact on growth and yield. Based on theoretical (Huhn 1985; Libby 1982) and empirical (Bentzer et al. 1988; Kleinschmit 1985) studies, a mixture of clones with 15-50 members provides sufficient stability against environmental interaction.

A critical component of this result is that the clones need to be diverse for adaptive traits (e.g., bud burst and bud set dates, resistance to pests) while they are selected for superior growth and yield. At the clone level, developmental homeostasis (ontogenetic self-regulation) (Lerner 1954) becomes much more important than with seedling-based programs, since clones are needed that display considerable homeostasis against environmental fluctuations. With a seedling-based stand of trees, the fate of an individual genotype is relatively unimportant because of the anticipated competition-induced mortality. However, with a clonal stand of trees (especially in some type of pure plot), the fate of an individual genotype is of paramount importance, since the clone is a single genotype. Mortality or loss of vigor by a single clone may cause loss of forest productivity over a large area.

7.4.2.2 Collection of cuttings and nursery strategy

To date, virtually all clonal planting stock arises from rooted cuttings. Tissue culture plantlets hold much promise, yet they remain too expensive to produce. Production systems are still under development. Cuttings are produced in some type of cutting orchard or stool bank from donor plants. The cuttings are collected during an optimum time window and set for rooting in some type of rooting system.

One basic decision to be made is whether to bulk the cuttings or to keep them separate by clone or groups of clones. An added production cost is incurred by maintaining the identity of all cuttings during the entire production process, yet separate identities of all clones provide the flexibility for the forest manager to deploy the rooted cuttings in a number of ways. The number of separate identities to be maintained can be reduced by grouping similar clones and then maintaining only the identity of the groups rather than the individual clones.

Root production on cuttings is strongly controlled by genetic effects (e.g., Foster et al. 1984; Wilcox and Farmer 1968); therefore, separating the clones within the greenhouse or nursery allows for specific cultural regimes and rooting times. For example, clones that root relatively quickly may begin to degrade before the slower rooting clones are ready to be removed from the rooting environment. Also, phenology of the clones may vary widely; and bulk mixes may suffer from suppression of the later flushing clones by the earlier flushing clones. The highly uniform behavior of ramets within a clone provides an opportunity to tailor the production system to produce rooted cuttings with uniformly-optimum characteristics. Rooted cuttings with low vigor, that result from poor cultural practices, will not express their true genetic potential, and hence genetic gain, in a forest stand.

7.4.2.3 Planting strategy

Clones can be deployed in the field as bulk mixes, groups of clones, or single clones. Furthermore, multiple clones can be deployed at one site either as an intimate mixture of all clones or as a mosaic of patches of groups of clones or pure clone patches (Libby 1987). Hence, a wide variety of deployment strategies is available once the number of clones per planting site has been determined.

Unfortunately, little information is available regarding yield advantages for any of the deployment strategies. Results reported by Markovic and Herpka (1986) with a 4-year-old eastern cottonwood (*Populus deltoides* Bartr.) study showed that the mean of a mixture of five clones exceeded the mean of the pure plot performance of the same clones for d.b.h., height, basal area, and volume; however, the means were not statistically different.

7.5 Growth and Yield Modeling to Predict Genetic Gain

Historically, genetic gain from tree improvement programs has been predicted by using formulas for either direct or part-record selection (Section 7.3) (Namkoong 1979). As noted above, this approach may be hampered by the lack of accurate parameter estimates (e.g., due to long rotations), severe competition effects on survival and frequency distributions, and the lack of a close relationship between a typical plot structure (i.e., size and configuration) in genetic tests and forest stand structure. Fortunately, the field of forest growth and yield modeling is advanced in general, and procedures are available that may greatly aid testing, selection, and genetic gain estimation. Tree breeders must first gain an understanding of the basic concepts in growth and yield modeling to integrate the procedures into tree improvement programs.

7.5.1 Variables commonly used in forest growth and yield modeling

During many decades of research, growth and yield modelers have found a few variables to be most useful in predicting current volume and future growth. These variables include, but are not limited to, the following (Clutter et al. 1983):

1. Volumes of individual trees, which are in turn a function of height, diameter at breast height (d.b.h.), and stem form,
2. Frequency distribution of the variables (e.g., diameter distribution),
3. Site index or average height of the dominant-codominant trees at a base age (e.g., 25 or 50 years),

4. Initial and current stand density per unit area (e.g., trees per acre),
5. Mortality trends over stand age,
6. Basal area per unit area (e.g., per acre),
7. Stand density versus average tree size.

Variables such as height and d.b.h. are commonly measured by tree breeders whereas others are easily calculated from planting records (e.g., mortality, initial tree density). Data are available in most tree improvement programs that could be readily used to predict future growth and yield of forest stands of genetically improved trees.

7.5.2 Predicting future growth and yield

Three basic approaches have been taken by growth and yield modelers. The first, and probably most common, is termed a *whole-stand model* in which the goal of the model is to simply predict the volume of a particular forest stand. The second main type of growth and yield model is termed a *diameter distribution model*, in which the parameters of the diameter distribution are predicted using stand attributes. The distribution is used to predict the stand volume. The third major type of model is termed an *individual-tree model* in which volumes of the individual trees are predicted and then summed to determine the stand volume. Each of the model types has strengths and weaknesses in terms of accuracy of prediction and output information.

7.5.2.1 Whole-stand models

Whole-stand models are particularly good for predicting total volume or merchantable volume, but little detailed information is available on individual tree attributes (Clutter et al. 1983). This model type might be useful to determine some future growth and yield of a pure family or pure clone stand, but it would not be particularly useful to predict the yield of a mixed family or mixed clone stand. These models are simple and generally require some or all of the following variables as input: current and future stand age, site index, basal area per unit area, initial number of trees, and number of surviving trees per unit area at the future age.

7.5.2.2 Diameter distribution models

The diameter distribution model is actually a hybridization of whole-stand and individual-tree modeling philosophies, and it predicts the diameter distribution and average height by diameter class. A frequency distribution, such as the Weibull distribution, is frequently used, although other distributions have also been utilized (Clutter et al. 1983). Average height, d.b.h. class, tree frequency per diameter class, and individual-tree volume equations are used to calculate tree volumes, which are then summed to predict total stand volume. Equations that generate diameter class frequencies and average heights generally require the stand age, site index, initial number of trees, and total number of surviving trees per unit area.

7.5.2.3 Individual-tree models

Individual-tree models are subdivided into distance-independent models and distance-dependent models. The major difference is whether the position and distance between trees in the stand must be determined. The distance between each subject tree and competing trees can then be determined for the model, and some type of competition index is usually computed. Individual-tree models may be more adaptable to projecting growth and yield in mixed genotype or mixed family stands because the identity of each tree can be maintained and used to independently model its growth and yield.

7.5.2.3.1 Distance-independent models

In distance-independent models, knowledge of the distance between trees is not needed. Instead, stand table data (e.g., number of trees/unit area and mean height by d.b.h. class), stand age, and site index are used to simulate the stand table data for a specified time in the future. The data are then compiled to predict volume by diameter class and total volume.

7.5.2.3.2 Distance-dependent models

Distance-dependent models generally require a "tree list," as well as other stand attributes. The tree list requires the following information for each tree for which growth is being projected: d.b.h., height, sometimes crown ratio or crown diameter, and a pair of X-Y coordinates specifying tree position in the stand. A competition index value is calculated for each tree and is used to determine the mortality probability and also the future growth for the tree. Individual tree volumes are calculated from the final predicted values and are compiled to estimate per unit area volume.

7.5.3 Modifying growth and yield models to accommodate genetic improvement

With the growth and yield modeling techniques available, the tree breeder and mensurationist must decide jointly which variables in the models should be modified and in what way to modify them. As can be seen from the previous discussion, there actually is a relatively small number of easily measured variables that mensurationists have found to be useful. Researchers in both disciplines must begin to target their research toward these well known variables.

7.5.3.1 Individual-tree volume

Volume equations serve as one of the essential elements of a growth and yield system, hence they must be accurate. Trees with the same height and d.b.h. can have differences in bole volume. These differences in actual volume are probably due to differences in bole taper which may arise from either environmental (Burkhart et al. 1981; Clutter et al. 1983) or genetic factors (Rockwood 1981). Thus, volume equations may need to be tailored to specific families or clones to provide accurate estimates. The basic forms of volume and taper equations are given by Clutter et al. (1983); therefore, clone- or family-specific parameters can be developed with sample information.

7.5.3.2 Frequency distributions

To date, few studies have tested for genetic differences in frequency distributions for forest trees. Buford and Burkhart (1985) found no significant differences in the variance of either 8-year-height or d.b.h. among 11 open-pollinated families and a woods-run checklot of loblolly pine. In their loblolly pine study, Janssen and Sprinz (1987) did not test for differences in variances between superior families and a checklot, but they did note a significant upward shift in the entire diameter distribution and a resultant improvement in the average maximum diameter of improved families versus an unimproved checklot.

Hypothetically, families and especially clones should display a smaller within-plot variance than bulk mixes or woods-run checklots, since there is less genetic variation. In addition, the distribution may differ between slower growing families and faster growing families when they are mixed in the same stand. The slower growing families may be forced into a left-skewed distribution after the onset of competition and before the onset of mortality, while the faster growing families may remain more normally distributed (see Section 7.2.1). Families may differ in their frequency distributions merely as a result of being in different development stages at the same age due to different growth rates or due to inherent difference in frequency distribution.

Diameter distribution growth and yield models are particularly sensitive to changes in frequency distributions. The frequency distribution, in effect, controls the rest of the growth and yield prediction, and it must be close to the actual distribution to reach an accurate prediction.

7.5.3.3 Site index

Site index and the related height-age functions are basic concepts in stand growth and yield estimation (Clutter et al. 1983). As described above, site index is a component of virtually all growth and yield models. The site index concept arose from many foresters' observations that sites differ widely in their basic productivity. One measure of this productivity is the dominant-codominant height of a stand at a base age (site index). The general relationship between stand height and age provides a background on which to base site index.

To date, most of the published research on genetic effects on stand growth and yield have used one of two height-age functions: the Schumacher function (e.g., Buford and Burkhart 1987; Nance and Wells 1981) or the Richards function (Knowe and Foster 1989; Sprinz 1987). The basic form of the Schumacher function is as follows:

$$\ln (H) = a + b (A^{-1}) + e \qquad \text{Eq. 7.3}$$

where,

H = height at any age, A
a = level parameter
b = slope or shape parameter
ln = base e logarithm
e = random error

The basic form of the Richards' function is:

$$H = a \{1 - \exp(-b A)\}^{c} + e \qquad \text{Eq. 7.4}$$

where,

a = asymptote parameter
b = rate parameter
c = shape parameter

Using the Schumacher function, Nance and Wells (1981) and Buford and Burkhart (1985, 1987) found family or seed source differences for 'a' but not for 'b'. Additionally, Knowe and Foster's (1989) results indicated significant differences among families for 'a' and 'b' but not for 'c' when using the Richards function. This suggests that, at least for loblolly pine, there may be consistent family differences in the level or asymptote of the height-age function with some evidence for differences in rate but not shape of the function.

Once the height-age function was characterized for each family and the woods run checklot in their study, Knowe and Foster (1989) used a stand simulator and separate height-age equations for each family and the checklot to predict future growth and yield. This approach did not require the part-record formula given in Section 7.3 with its inherent weaknesses. Use of the stand simulator provided reasonable estimates of genetic gain in height (within 2% of prediction for ages 2 to 15 years) and per acre volume (within 3% of prediction for ages 5 to 15 years using the unthinned option) compared to estimates from the traditional prediction equation. Although the data set only extended to age 15, the simulator provided estimates through age 25 (rotation age). This approach could actually replace the use of indirect selection for juvenile traits and correlated response in rotation length traits, which is commonly used in tree improvement.

7.5.3.4 Stand density versus average tree size

The general applicability of the "- 3/2 Power Law" across many tree species (Nance et al. 1987) indicates that there may be little difference among genetic entities (e.g., seed sources, families, or clones) within a species for the slope (or rate of change) of the log of average tree size versus the log of stand density. Two studies have been published that relate to this issue with forest trees; and in both cases, the authors found that the intercepts, K value (Nance et al. 1987), of the linear relationship differed significantly among seed sources although the slopes of the lines were not significantly different (Buford 1989; Schmidtling 1988). Hence, the available evidence suggests that at least seed sources may vary in their carrying capacity. Furthermore, both studies showed a significant positive relationship between site index of the seed source and the K coefficient of the seed source. Nance et al. (1987) outlined a number of testable hypotheses regarding genetic effects on the size-density relationship. These relationships are basic to the understanding of how genetic improvement will manifest itself in stand productivity.

7.5.3.5 Mortality trends over stand age

Genetic tests, which generally are carefully cultured, may suffer relatively low mortality rates; therefore, mortality trends over stand age would provide an unrealistic estimate compared to those in operational plantations. Mortality related to competition is a basic process in stand development; hence mortality functions are an integral part of virtually all growth and yield models (Clutter et al. 1983). In their stand simulation, Knowe and Foster (1989) found a consistent under-prediction of survival with an operationally-based model compared with actual progeny test results. Therefore, researchers should expect at least some discrepancy between actual volume, derived from progeny tests, and volume predicted from stand simulators, that are based on operational stand development. Stand simulators still are probably more realistic for expected operational results. Tree breeders must conduct operational scale trials of genetically improved material to determine whether differences in survival detected in genetic tests are realistic. Competition-derived mortality is inevitable in a forest stand; yet genetic entities may differ in the number of surviving trees per acre (stand density) at a given tree size (Buford 1989; Schmidtling 1988). Mortality will still increase with stand age, even in genetically improved stands, but the mortality functions may differ significantly. If survival can be genetically improved, then the mortality functions within growth and yield models must be changed to reflect the improved survival.

7.6 Verifying Realized Genetic Gain

The task of verifying realized genetic gain is arduous; yet it must be accomplished to serve as a feedback mechanism to improve the effectiveness of the tree improvement program and to provide economic justification for the program. Several practical steps can be taken to allow for the future assessment of realized genetic gain.

Accurate records must be kept regarding the genetic composition of seed used to establish each stand. These records should include identity of the seed orchard where the seed arose, parental clones in the orchard, seed collection year, year the orchard was last rogued, which clones were rogued, etc. This information should be kept in an accessible form for the tree breeder and the various forest managers. As the forest managers become familiar with the behavior of the various families or clones, they will become more active in deploying them on their management units.

Appropriate checklots should be incorporated into the stands established with genetically improved planting stock. The checklots should represent unimproved populations; the deviation from which would serve as an estimate of realized genetic gain. The tree breeder should choose checklots based both on the local natural population and the best available unimproved source. Enough seed must be collected and stored so that the same checklots can be used extensively for at least ten years. In an operational tree improvement program, it is imperative that realized genetic gain be measured against the same standard for several generations.

A variety of methods is available for incorporating the checklots. To be successful, the measurement plot(s) should be located at random (with respect to environmental quality) within the entire plantation and should be surrounded by sufficient border trees so that the measurement trees are not influenced by adjoining open space or other treatments. Each plot should be a minimum of 0.75 acre in size so that at least 100 dominant-codominant trees will be present at rotation length. It is not necessary that all plantations include a sample plot of the checklot, but the range of environmental quality must be sampled to obtain sufficient accuracy for the realized genetic gain estimate.

Mensurational techniques are available to cruise both the improved plantations and the sample plots of the checklots in order to estimate per unit area yield and product quality. Most forestry organizations have customized inventory and growth and yield estimation procedures.

7.7 Conclusion

Tree improvement is a young area of applied genetics as compared with other plant and animal species. Few examples are available in the entire world in which even a single generation of improvement has been followed with an entire rotation length of trees leading to estimates of realized genetic gain. Quantitative genetic theory has clearly outpaced empirical results. All of the factors listed in this chapter (as well as others) must be investigated in order to either alter tree improvement practices or alter estimated genetic gain.

Literature Cited

Adams, W.T., and Birkes, D.S. 1989. Mating patterns in seed orchards. 20th South. For. Tree Imp. Conf., Charleston, South Carolina. pp. 75-86.

Barnett, J.P., and Dunlap, J.R. 1982. Sorting loblolly pine orchard seeds by size for containerized seedling production. *S. J. Appl. For.* 6:112-115.

Barnett, J.P., and McLemore, B.F. 1984. Germination speed as a predictor of nursery seedling performance. *S. J. Appl. For.* 8:157-162.

Bentzer, B.G., Foster, G.S., Hellberg, A.R., and Podzorski, A.C. 1988. Genotype x environment interaction in Norway spruce involving three levels of genetic control: seed source, clone mixture, and clone. *Can. J. For. Res.* 18:1172-1181.

Buford, M.A. 1989. Mean stem size and total volume development of various loblolly pine seed sources planted at one location. *Can. J. For. Res.* 19:396-400.

Buford, M.A., and Burkhart, H.E. 1985. Dynamics of improved loblolly pine plantations and the implications for modeling growth of improved stands. 18th South. For. Tree Imp. Conf., Long Beach, Mississippi. pp. 170-177.

Buford, M.A., and Burkhart, H.E. 1987. Genetic improvement effects on growth and yield of loblolly pine plantations. *For. Sci.* 33:707-724.

Bulmer, M.G. 1985. *The Mathematical Theory of Quantitative Genetics.* Clarendon Press, Oxford.

Burkhart, H.E., Cao, Q.V., and Ware, K.D. 1981. A comparison of growth and yield prediction models for loblolly pine. Pub. No. FWS-2-81, School of For. and Wildlife Resour., Virginia Polytechnic Institute and State Univ., Blacksburg, Virginia.

Cannell, M.G.R. 1982. 'Crop' and 'isolation' ideotypes: evidence for progeny differences in nursery-grown Picea sitchensis. *Silvae Genetica* 31:60-66.

Cannell, M.G.R. 1984. Competition and selection for yield: a perspective from forestry. In *Efficiency of Plant Breeding* (W. Lange, A.C. Zeven and N.G. Hogenboom, eds), pp. 28-34. Proc. 10th Congress of Eucarpia, Pudoc, Wageningen, Netherlands.

Cannell, M.G.R., Bridgwater, F.E., and Greenwood, M.S. 1978. Seedling growth rates, water stress responses, and root-shoot relationships related to eight-year volumes among families of *Pinus taeda* L. *Silvae Genetica* 27:237-248.

Clutter, J.L., Fortson, J.C., Pienaar, L.V., Brister, G.H., and Bailey, R.L. 1983. *Timber Management: A Quantitative Approach.* John Wiley & Sons, New York.

Falconer, D.S. 1981. *Introduction to Quantitative Genetics, 2nd ed.* Longman, New York.

Ford, E.D. 1975. Competition and stand structure in some even-aged plant monocultures. *J. Ecol.* 63:311-333.

Foster, G.S. 1986. Provenance variation of eastern cottonwood in the lower Mississippi Valley. *Silvae Genetica* 35:32-38.

Foster, G.S. 1989. Inter-genotypic competition in forest trees and its impact on realized gain from family selection. 20th South. For. Tree Imp. Conf., Charleston, South Carolina. pp. 21-35.

Foster, G.S., Campbell, R.K., and Adams, W.T. 1984. Heritability, gain, and C effects in rooting western hemlock cuttings. *Can. J. For. Res.* 14:628-638.

Gladstone, W.T., Bean, P.M., Hughes, J.H., and Talbert, C.B. 1987. A challenge for tree improvement. 19th South. For. Tree Imp. Conf., College Station, Texas. pp. 1-8.

Hallauer, A.R., and Miranda, J.B. 1981. *Quantitative Genetics in Maize Breeding.* Iowa State University Press, Ames.

Hühn, M. 1985. Theoretical studies on the necessary number of components in mixtures. I. number of components and yield stability. *Theor. Appl. Genet.* 70:383-389.

Janssen, J.E., and Sprinz, P.T. 1987. Modeling distributions of stem characteristics of genetically improved loblolly pine. 19th South. Forest Tree Imp. Conf., College Station, Texas. pp. 367-375.

Kleinschmit, J. 1985. Concepts and experiences in clonal plantations of conifers. In *Clonal Forestry: It's Impact on Tree Improvement and Our Future Forests.* (L. Zsuffa, R.M. Rauter, and C.W. Yeatman, eds). 1983. 19th Can. Tree Imp. Assn. Mtg., pp. 26-56. Canadian Forestry Service, Chalk River.

Knowe, S.A., and Foster, G.S. 1989. Application of growth models for simulating genetic gain of loblolly pine. *For. Sci.* 35:211-228.

Lambeth, C.C., Gladstone, W.T., and R.W. Stonecypher. 1983. Statistical efficiency of row and noncontiguous family plots in genetic tests of loblolly pine. *Silvae Genetica* 32:24-28.

Lerner, I.M. 1954. *Genetic Homeostasis.* Oliver and Boyd, London.

Lerner, I.M. 1958. *The Genetic Basis of Selection.* John Wiley & Sons, Inc. New York. 298 p.

Libby, W.J. 1982. What is a safe number of clones per plantation. In *Resistance to Diseases and Pests in Forest Trees* (H.M. Heybroek, B.R. Stephan, K. von Weissenberg, eds). 1980. pp. 342-360. Centre for Agricultural Publishing and Documentation, Wageningen.

Libby, W.J. 1987. Testing for clonal forestry. *Annales Forestales* 13:69-75.

Libby, W.J., and Cockerham, C.C. 1980. Random and non-contiguous plots in interlocking field layouts. *Silvae Genetica* 29:183-190.

Loo-Dinkins, J.A., and Tauer, C.G. 1987. Statistical efficiency of six progeny test field designs on three loblolly pine (*Pinus taeda* L.) site types. *Can. J. For. Res.* 17:1066-1070.

Marković, J., and Herpka, I. 1986. Plantations in short rotations. In *Poplars and Willows in Yugoslavia* (V. Guzina, ed), pp. 182-197. Poplar Research Institute, Novi Sad.

Martin, G.A., Glazener, E.W., and Blow W.L. 1953. Efficiency of selection for broiler growth at various ages. *Poultry Sci.* 32:716-720.

Mohler, C.L., Marks, P.L., and Sprugel, D.G. 1978. Stand structure and allometry of trees during self-thinning of pure stands. *J. Ecol.* 66:599-614.

Namkoong, G. 1979. *Introduction to Quantitative Genetics in Forestry.* USDA Forest Service, Tech. Bull. No. 1588.

Nance, W.L., and Wells, O.O. 1981. Estimating volume potential in genetic tests using growth and yield models. 16th South. For. Tree Imp. Conf., Blacksburg, Virginia. pp. 39-46.

Nance, W.L., Land, Jr., S.B., and Daniels, R.F. 1983. Concepts for analysis of intergenotypic competition in forest trees. 17th South. Forest Tree Imp. Conf., Athens, Georgia. pp. 131-145.

Nance, W.L., McCutchan, B.G., Talbert, C.B., Buford, M.A., Foster, G.S., and P. Sprinz. 1987. Experimental approaches for evaluating genetic effects on stand growth and yield. In *Statistical Considerations in Genetic Testing of Forest Trees.* pp. 22-39. South. Coop. Ser. Bull. 324, University of Florida, Gainesville.

Nanson, A. 1970. Juvenile and correlated trait selection and its effect on selection programs. 2nd Meet. Work. Group on Quant. Genet., Sect. 22, IUFRO, 1969. pp. 17-26. USDA For. Ser., South. For. Exp. Stn., New Orleans.

Rockwood, D.L. 1981. Volume prediction for genetically improved slash pine trees. 17 p. Agri. Exp. Sta. Bull. 819, Univ. Florida, Gainesville.

Schmidtling, R.C. 1988. Racial variation in self-thinning trajectories in loblolly pine. In *Forest Growth Modelling and Prediction* (A.R. Ek, S.R. Shifley, and T.E. Burk, eds), pp. 611-618. USDA For. Ser., North Central For. Exp. Stn. Gen. Tech. Rep. NC-120.

Silen, R., and Osterhaus, C. 1979. Reduction of genetic base by sizing of bulked Douglas-fir seed lots. *Tree Planters' Notes* 30:24-30.

Sniezko, R.A., and Zobel, B.J. 1988. Seedling height and diameter variation of various degrees of inbred and outcross progenies of loblolly pine. *Silvae Genetica* 37:50-60.

Sorensen, F.C. 1971. Estimate of self fertility in coastal Douglas-fir from inbreeding studies. *Silvae Genetica* 20:115-120.

Sprinz, P.T. 1987. Effects of genetically improved stands on growth and yield principles. 19th South. For. Tree Imp. Conf., College Station, Texas. pp. 338-348.

Squillace, A.E. 1974. Average genetic correlations among offspring from open-pollinated forest trees. *Silvae Genetica* 23:149-156.

Squillace, A.E., and Gansel, C.R. 1974. Juvenile:mature correlations in slash pine. *For. Sci.* 20:225-229.

Squillace, A.E., and Kraus, J.F. 1962. Effects of inbreeding on seed yield, germination, rate of germination, and seedling growth in slash pine. In *Proc. Forest Genetics Workshop*, pp. 59-63. Tree Imp. Comm. of Soc. Amer. For. and South. For. Tree Imp. Comm. Sponsored Pub. No. 22.

Stern, K., and Roche, L. 1974. *Genetics of Forest Ecosystems.* Springer-Verlag, New York.

Trenbath, B.R. 1974. Biomass productivity of mixtures. *Adv. Agron.* 26:177-210.

Turner, H.N., and Young, S.S.Y. 1969. *Quantitative Genetics in Sheep Breeding.* Macmillan of Australia, South Melbourne.

Van Vleck, L.D., and Doolittle, D.P. 1964. Genetic parameters of monthly egg production in the Cornell controls. *Poultry Sci.* 43:560-567.

Van Vleck, L.D., and Henderson, C.R. 1961. Estimates of genetic parameters of some functions of part lactation milk records. *J. Dairy Sci.* 44:1073-1084.

Wilcox, J.R., and Farmer, R.E. 1968. Heritability and C effects in early root growth of eastern cottonwood cuttings. *Heredity* 23:239-245.

Wright, J.W., and Freeland, F.D. 1960. Plot size and experimental efficiency in forest genetic research. Michigan State University Agri. Exp. Stn. Tech. Bull. 280. 28 p.

Chapter 8
Quantitative Approaches to Decision-Making in Forest Genetics Programs

S.T. Friedman[1]

Turning "seat of the pants" decision-making into a quantitative evaluation is the heart of quantitative genetics. We apply the tools of mathematics to the tasks of selecting, breeding and deploying forest trees. Yet when it comes to using economic criteria to quantify the results of our programs, or to help us make decisions within our programs, forest geneticists have had a tendency to revert to heuristic ways of making decisions. In this chapter I will refer to genetic resource management (GRM) programs, by which I mean programs that have the responsibility of securing stocking at desired densities of desired species. The source of this stocking should be dependable, genetically variable, genetically adapted to the site, and may be selected for some desirable traits. My use of the term includes activities that range from the most extensive (natural regeneration and/or wild stand collections from local sources) to the most intensive (genetic engineering), which would enable us to examine a broader

---

[1]*Research Coordinator, RPA, USDA Forest Service, Washington, D.C.*

L. Fins et al. (eds.), Handbook of Quantitative Forest Genetics, 270–312.

range of options than commonly considered in "tree improvement" analyses.

While I argue in this chapter that economic criteria should be considered in evaluation of GRM programs, arguments can also be made against the sole use of economic criteria in decision-making. Certainly not all GRM program decisions should be made on a strictly economic basis. Gene-resource management decisions may be justified for other reasons such as:

1) *Genetic conservation/maintenance of variation*; e.g., should more lodgepole trees be selected in areas in which all the current selections have been killed by mountain pine beetle?

2) *Seed production*; e.g., should a grafted larch orchard be established solely because of difficulty obtaining seed in wild stands?

3) *Information needed to manage the program efficiently*; e.g., what are the genetic variances and covariances among traits necessary to make decisions that increase height growth without decreasing cold hardiness?

Despite the fact that many decisions may involve other than strictly economic criteria, certainly consideration of the costs to achieve a given objective - for example, gene conservation- is not out of place. The benefits to be balanced against the costs in GRM programs are generally considered in terms of genetic gain in a desired trait. Retention of genetic variability is perhaps more important, but is also more difficult to quantify. So analysis falls into two main categories: 1) accomplishing stewardship goals at the desired quality at least cost (cost-effectiveness analysis), and 2) investments in tree improvement which are expected to yield a return (economic efficiency analysis).

Analysis of programs from an economic point of view can indicate which benefits are important to profitability, which costs are affecting profitability, and which investments in gaining more information are more likely to pay off. Analysis identifies areas where further research is necessary (van Buijtenen *et al.* 1975).

"The obvious goal is to make a decision that will stand the test of time by being as wise and efficient tomorrow as it is today. The value of an analysis is not to foretell the occurrence of unknown events, but to identify the vulnerability of management options to uncertain events, and to suggest alternatives that are flexible and ingenious enough to be efficient under a wide array of future outcomes" (Connaughton 1986).

8.1 Uses for Economic Criteria

Economic evaluation of alternatives can occur at several levels within a GRM program. First, decisions can be made on program levels. For example, Ford *et al.* (1983) compared 1) control of seed source; 2) establishment of seed production areas; 3) development of first generation seed orchards, and 4) creation of advanced generation seed orchards. Economics can also be used to choose traits to measure, for both direct and indirect selection. Examining a wide range of program levels and alternative testing strategies, (e.g., mass selection, seed production areas, greenhouse and short-term tests), in addition to the standard seed orchard and progeny tests or clonal approach, can lead to a broader understanding of the costs and benefits of GRM programs.

Another use for economic evaluation is to decide on the appropriate management intensity for seed production areas, seed orchards, mass selection plantations, clone banks, selected trees, and progeny tests. Within each type of planting, there exists a continuum of management options from extensive to intensive. With different objectives, different problems, and/or different values for each species in each zone, the economically optimal strategy might be expected to vary for each site. Economic analysis, including consideration of risk, can often bring a structure to bear on decisions which can help to elucidate an appropriate strategy for the management of a given planting.

Given that the desired management alternatives for each zone are known, economics can be used to establish priorities. In the vast majority of cases, either time or budget limits what might be accomplished. The question often arises: "On which of these many projects should I begin? Where would I get the greatest pay-off?". These types of questions occur both at the long term budgeting level, and also on the day-to-day level. For example, the progeny test data that should be analyzed first, or the block of the seed orchard that gets most careful monitoring, should be those in which the expected benefits for the work performed would be the greatest.

In the examples below, one simple way to answer some of the above questions will be illustrated. More sophisticated techniques, such as overall program optimization (Paques 1984), also exist. However, optimization supposes a degree of certainty about goals and desired outputs which may or may not reflect the attitudes of a given organization.

Finally, the use of economic valuation of traits is essential for selecting trees for the greatest financial gain using selection indices or best linear prediction. Some papers that describe the use of economic criteria include Bridgwater and Squillace (1986), and Bridgwater and Stonecypher (1979).

8.2 Economic Analysis Example

The objective of this section is to provide a step-by-step guide to conducting a simple economic analysis, with examples. This section presupposes a familiarity with economic criteria such as benefit cost ratio and net present value (NPV). For a review of these criteria and their applications, see Clutter *et al.* (1983).

8.2.1 Information needed for analysis

1. *Organizational Objectives* Probably the most important factor in any analysis is to know the objectives of your organization, and be able to explicitly state them. Some possible objectives for private forest companies might include: 1) maximize NPV or 2) reforest to meet legal requirements at lowest cost. On the other hand, public agencies have the responsibility for both land stewardship, which involves maintaining diversity and adaptation, and for making biologically and

financially responsible investments by changing population means and/or variances of traits for which economic benefits outweigh costs.

2. *Accounting: Research vs. Operations* In some organizations, progeny testing is considered to be research, while in some others it is placed in operations. Most geneticists have limited power to change the accounting, or responsibilities, within an organization. Yet, it is clear that if progeny testing is considered to be research (and separately funded), the results of an economic analysis strictly from the operational viewpoint would very likely be different than if the returns from the orchard also had to cover existing progeny test costs. McKenney *et al.* (1988) consider selection and progeny testing to be research and development costs due to the permanent nature of the gains, so that associated costs are ignored in an operational analysis. It has been suggested (Hansen[2], pers. comm.) that there might be a more objective basis for making decisions on whether some GRM activity should be a function of research or operations. This would be based on whether or not expected results are incorporated into planning for the future. While this may seem to be only a methodological concern, funding for GRM programs may depend on the accounting stance of the organization.

3. *Accounting: Genetics vs. Reforestation* Similarly, many organizations have separate accounting for genetics and reforestation, whereas in others, they are integrated. But the boundaries may not be clear. For example, fertilizing a seed production area might be funded from

---

[2]*C. Hansen, Mount Baker-Snoqualmie National Forest, Seattle, WA.*

genetics, because it is a seed production area, and fertilization will increase the quantity of genetically improved seed. But once the genetics work has been done (say, roguing the seed production area), perhaps all the work which leads to seed production should be funded from seed procurement. For small organizations, there will probably be no such divisions, but in larger organizations they may occur. The end result is that savings from, say, establishing a seed production area, may accrue to a seed collection account. While the source of funding is not generally relevant to the economics of the program, both costs and benefits to the organization may be hidden by the way funding is arranged. If an economic analysis is to be done, all relevant costs and benefits, to whatever part of the organization, must be examined. That's why it is difficult to do just a "tree improvement" economic analysis. Rather, it is important to do a whole-organizational "regeneration establishment" economic analysis. Yet when different functions get involved, an analysis can often be delayed to the point where it is never accomplished. So for each analysis, there are often trade-offs to be made between the scale of the analysis and the desire for timely completion.

4. *Economic Parameters* These include the discount rate, stumpage prices (current and future), and logging costs (current and future). Parameters relevant to cost and price projections include inflation and respective real rates of increase or decrease. It is important both to use the figures currently in use by the rest of the organization, and to use a range to test sensitivity of the decision to the set of parameter values chosen.

5. *Genetic and Seed Orchard Factors* These are the common questions of how much gain in what trait is obtained using different options, when the improved propagules are available and how long seed orchards, cutting hedges, etc, can produce propagules, and the costs associated with GRM activities. While many analyses are based on improved timber or pulpwood quality and/or quantity, other effects can be considered. These might include reduction in herbicide costs due to faster growth of young trees, lower nursery costs due to increased germination rate of orchard seed, or other savings. Though these are common questions, getting the answers may be much more difficult, especially for expected gain and its timing. Therefore, again, it may be desirable to use a range of values to test the sensitivity of the analysis to the assumptions used.

6. *Seed Orchard Market* Many studies assume that extra seed can be sold or planted, thereby increasing value with extra seed production. Perhaps once an organization's seed needs are filled there will be no additional market for seed for a zone, therefore no extra value.

7. *Growing Trees in Stands* Added value from the use of improved stock is only meaningful in the context of a given silvicultural prescription, site productivity class, etc. While geneticists may leave these details in the hands of the silviculturist or biometrician working with them on the analysis, some acquaintance with these factors is essential for good decision-making in GRM programs. In particular, quantification of acres to be reforested and seed needs (or a range of values) is very important information for an effective analysis.

8. *Externalities (and indirect benefits)* These are costs and benefits of GRM programs not included directly in the above decisions. These might include: fewer applications of herbicide leads to positive public relations, or fast growing tree plantations capture a site quickly after a fire and reduce erosion. From the standpoint of biodiversity, there are species such as sugar pine which might die out from a stand without the resistant trees provided by GRM investments. The lack of sugar pine could lead to a reduction in the populations of associated species to which sugar pine is the host.

It is important to mention externalities which have recently been discussed in the ecological literature. That is that through genetic improvement, trees would become more susceptible to some unknown risk factor. For example, decreased wood specific gravity could occur due to selection for some other trait, if wood specific gravity were not monitored. Or increased wood specific gravity could lead to increased snow breakage. Or, as in Rehfeldt (1989), selection for height growth alone could change dates of budburst or budset, leading to increased frost damage, if knowledge of phenology is not taken into account in selection for increased height. Our arguments as geneticists are that we account for those potential changes by measuring intercorrelations among adaptive traits, or by working within a zone with consistent adaptation. Other arguments are that GRM programs can decrease genetic variation within populations and/or decrease the variation among populations, (or change the architecture of tree populations) which could lead to reduced capacity to respond to stress. These would be externalities of the GRM programs if they occurred. Certainly using only natural regeneration, or planting only orchard seed might be expected to have different effects on diversity and structure, but at present, little information is available on the importance of these extreme alternatives. So pointing out that genetic diversity has a value, albeit difficult or impossible to quantify, may be as much as we can do at the present time. While this may not be important in choosing whether to artificially regenerate (e.g., when natural regeneration fails, and species diversity goals specify some reforestation), considerations of the possible risks can certainly affect choices of options within GRM programs.

9. *Opportunity Costs* These relate to direct costs in that the investment represents an allocation of resources (land, capital, labor) which foregoes some other opportunity. The opportunity cost is the value of the next best alternative foregone. In NPV, benefit-cost analysis, this cost is represented by the discount rate, sometimes called the alternative rate of return. A discussion of opportunity costs and externalities in silviculture can be found in McDivitt and Hansen (1983).

8.2.2 Six steps to a simple economic analysis

The objective of developing this simplified approach is to take the first step of discounting (discounting costs and benefits throughout the life of a prescription) and separate these figures from the rest of the analysis. This has the advantage that the results of the growth model runs, stumpage values, and logging costs can be arrived at according to standard practices, and the geneticist can translate the results into a single number (marginal value of genetic gain per unit area or MVG/A) and then calculate the values of program options. The

MVG/A can then easily be compared with other silvicultural investments. This separation also enables seed orchard managers and others to be able to arrive at crude estimates of the economic value of various manipulations using a pocket calculator. The simplicity requires at least two fairly restrictive assumptions, however. The first is that the number of acres to be planted per year is constant over the time period analyzed (e.g., always 200 per year, not 100 for the first 20 years, then 50 for the next 10 and so on). The second assumption is that there is no real price increase or that the real price increase stops before the first volume is removed. If the real price increase stops before the first volume is removed, the real price increases may be treated as constants. With a continuing real price increase, the actual years of harvesting of improved trees must be known and, as with changes in acreages planted, more computational work must be done either with a computer program or a spreadsheet (see Hansen 1990).

The following fictitious example compares two alternatives, 1) wild stand collections and 2) seed orchards with progeny tests on the Bovine Breeding Zone of the Yellowbelly Forest. The goal is to develop a break-even number of acres planted for which the benefits of establishing a seed orchard and progeny tests equal the costs.

It was noted throughout the last 20 years on the Yellowbelly Forest that planned acres to be planted per year changed a great deal. By conducting a break-even analysis by acres, if planted acres went above or below the break-even figure, action could be taken to decrease or increase the investments in the program.

Step 1. Decide on which variables will be affected by GRM investments, when and for how long the improved material will be used. The simplest would be to consider traits such as height or height and diameter. Survival, especially if it could lead to reduced replanting costs, would also be relatively easy to consider. Pest resistance, form, and other traits could be included as well. The gains for the different traits under different options must be estimated, and also the times when the propagules with different amounts of gain will be available.

In the Bovine example, the only trait considered is height growth. The seed orchard is assumed to produce propagules that are first planted in year 16 from establishment of the orchard. The orchard produces these propagules for 30 years, from ages 16 to 45. Since the investment begins at the establishment of the seed orchard, values need to be discounted twice, once from throughout the rotation back to time of planting, and once from planting of each reforestation acre to the establishment of the orchard. This will be shown in a later formula. For now, the orchard side of the discounting can be shown in gain/time diagrams (Figures 8.1 and 8.2).

Step 2. Calculate the present values of the costs of each option. Calculate the present value of the costs of the different alternatives to be compared. This can be done using Equation 8.1:

$$PVC=\sum_{t=0}^{n}\frac{C_t}{(1+i)^t} \qquad \textbf{Eq. 8.1}$$

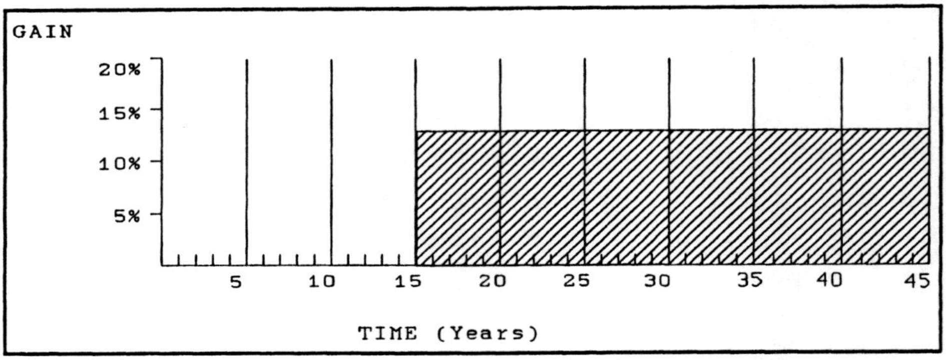

Figure 8.1. This gain/time diagram shows that from years 16-45, seed orchard stock is being planted, which has an expected gain of 13% in height. This type of diagram shows that the value stream from years 16-45 needs to be discounted to time 0 (if costs are calculated from time 0). Years, beginning with year 1, are represented by the spaces between the marks along the x axis.

where i equals interest rate, PVC is the present value of the costs, C is the value of the costs at time t, and the discounted costs are summed from time 0 to the end of the project. Most people are familiar with the costs of different kinds of activities. Discounting can more readily be done using a computer program or a spreadsheet. Costs associated with seed orchards usually fall into categories such as site selection, site preparation, facilities, orchard establishment, orchard maintenance and orchard roguing. Tree selection can be charged to the seed orchard, to the seed orchard and progeny test, or to recurrent phenotypic selection programs. Progeny testing involves site selection, preparation, design and layout, planting, protection, measuring and analysis. A spreadsheet including these costs can be found in Hansen (1990).

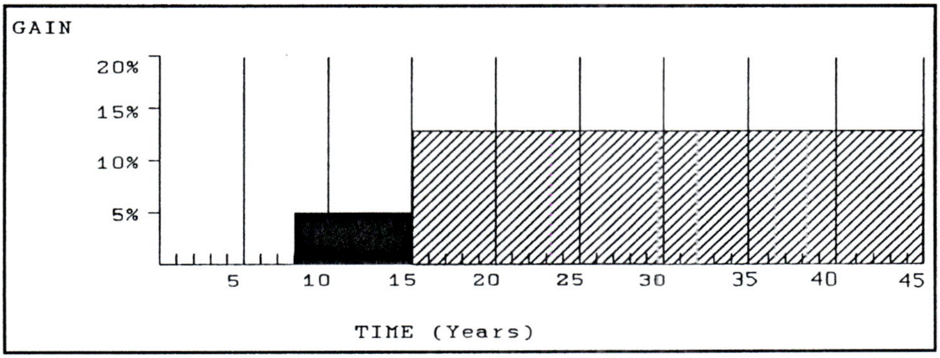

Figure 8.2. This gain/time diagram shows an assumption of 5% gain in height from open-pollinated seed collected in the woods from the top 25% of parents based on 5-year measurements in a progeny test. This assumes that the orchard and the progeny tests were planted at the same time, that it takes two years to collect from the parents, and two to grow seedlings in the nursery. The gain from years 9-15 due to the planting of open-pollinated seed from the tested parents is shown in the shaded area.

In the Bovine example, the interest rate was assumed to be 4%, the same rate used in the Yellowbelly Forest Plan. $300,000 was assumed to be the present value (discounted to present) of the stream of costs due to establishment and maintenance seed orchards and progeny tests.

Step 3. Calculate the present value of the benefits of the alternatives. These can fall into two major areas: first, **savings** from standard practices due to the availability of tree improvement stock (e.g., collecting from seed orchards may yield savings compared to collecting from wild stands), and second, **additional** monetary **value** from products. This can be caused by obtaining the same quantity and/or quality of product in a shorter period of time, or a larger quantity and/or quality of product in the same period of time.

Calculating additional value due to improved stock can range from the relative simplicity of a gain in height converted to an increase in site index, to gains in height, diameter, survival, and form, depicted by differences in site index, diameter multiplier, mortality function, and taper equation. More information on this subject is available in Foster's chapter in this volume. Currently the lack of data from yield trials make this estimation one of the weaker areas of the analysis process.

In the Bovine example, we assume that a tested seed orchard will produce trees that have a 13% increase in height. This was derived from some data on the species, some genetics theory, and some heuristic adjustment downward for risk and the fact that the data are from progeny tests and not from yield trials. Probably it would be wiser to bracket the range at 5-20 percent, perhaps in 5 percent increments, so that a sensitivity analysis could be conducted on these estimates of yield gain. We also assume that, with no progeny tests and no seed orchards, gain would be zero. So in this analysis we are comparing a 13% increase in height with the base level. This is depicted in the growth model by using site index 67 for unimproved and 76 for improved stands.

Step 3a. Divide the breeding/seed/deployment zone into areas that differ for management and/or biological reasons. Naturally, this assumes that the analysis is done on a given species and zone. There are advantages to doing analyses on a local, rather than say, state/province/program wide basis. One is that changes in management locally are instantly translatable into changes in programs. This provides a great deal more flexibility, and greater integration with other silvicultural options. Some examples of breakdowns within a zone would be ponderosa pine and mixed conifer, different site classes, and land suitable for different harvesting systems (tractor vs. cable). Often a clue to a useful breakdown is that other silvicultural analyses are done with the same categories.

In the Bovine example, runs were done separately for ponderosa pine and mixed conifer, and for tractor and cable ground within each type. The proportions were: mixed conifer tractor .26; mixed conifer cable .14; ponderosa pine tractor .45; and ponderosa pine cable .15.

Step 3b. Do growth model runs for each category in Step 3a for each genetic gain value in Step 1. Prescriptions to be input into the growth models can be those currently in use or the range of possibilities imagined in the future. The prescription with the highest NPV can be selected if there are no other constraints. It is possible that optimum economic rotation could be shortened (McKenney *et al.* 1988) and if so, this could be included.

In the Bovine example, UGROWM, a fictitious stand model that has site index as a driver, was used to project outputs and values for the two alternatives at various rotation ages that were determined by the manager. These can be found in Table 8.1. Also in the Bovine example, we assume that we are going to do one rotation of timber management, therefore we will simply stop the shorter rotation, rather than beginning another rotation using a multiple rotation Faustmann or equivalent annual income approach. Although this will underestimate gains, we have little confidence in predicting management regimes on these ownerships beyond 150 years.

Step 3c. Calculate incremental gains in NPV due to GRM investment for each silvicultural prescription/site/logging system combination on a per acre basis. As stated above, calculating per acre values enables people to compare increases in value due to genetic improvements with other investments. This may be useful for doing pocket calculator analyses in the field during discussions of program options. Note that an interest rate is needed here to calculate the NPV of each prescription. If the interest rate is changed for sensitivity analyses, this step needs to be repeated for each interest rate.

The net values (stumpage minus logging costs) for prescriptions with the highest NPV are shown in Table 8.1 for three points in time and four timber type/logging combinations. In addition to the extra volume, or time gained (see the category mixed conifer tractor, where the time to the first commercial entry is shortened by ten years), savings can also be added in here which might accrue due to decreased management costs on a plantation. Finally, the difference between genetic sources of stock is obtained by subtracting the NPV from the lower level of genetic stock from the NPV of the higher.

Table 8.1 Prescriptions with the highest net present value and marginal value of genetic gain per acre for four timber type/logging cost combinations for the Bovine Breeding Zone.

Timber/Logging Type	Seed Orchard (SI=76) Age[1]	$/Acre[2]	Unimproved (SI=67) Age	$/Acre	MVG/A[3]
Mixed Conifer	40	322	50	143	
Tractor	60	419	70	511	
	80	2603	90	2995	
NPV[4]		220		141	79
Mixed Conifer	60	274	60	170	
Cable	90	3274	90	2708	
NPV		122		96	26
Ponderosa Pine	50	272	50	155	
Tractor	70	328	70	314	
	90	2588	90	2124	
NPV		135		104	31
Ponderosa Pine	50	204	50	112	
Cable	70	257	70	244	
	90	2196	90	1784	
NPV		110		84	26
Combined					42[5]

[1]Age = age at harvest
[2]$/Acre = net value per acre (stumpage - logging costs)
[3]Marginal value due to genetics/acre
[4]NPV's using 4% interest rate and no real price increase in stumpage (values at each age discounted to present and summed)
[5]MVG/A weighted by number of acres in each timber/logging combination

This gives a figure which this author terms the marginal value of genetic gain per unit area or MVG/A for a given genetic source, silvicultural prescription, site class, logging system, stumpage value and interest rate. The MVG/A can then be used for field analyses of options, as in Boxes 8.1 and 8.2.

Note also that this NPV includes the costs and benefits of a silvicultural prescription with different genetic options. The costs of the program itself, although calculated in step 2, are not yet included. That will occur in a later step. The goal of this step is simply to get an idea, on an area basis, of how much the value of a prescription might be increased by a investment in genetic improvement.

Step 3d. Weight together different biological and management categories to get average value for the entire zone. The MVG/A's (numbers in the furthest right hand column of Table 8.1) are multiplied by their proportions to get an average value for the zone. In the example presented in Table 8.1, the value was obtained as follows: (.26 x \$79) + (.14 x \$26) + (.45 x \$31) + (.15 x \$26) = \$42, the average MVG/A for all acre. Again, the MVG/A is the additional value due to genetic investments on an acre, discounted to the time the acre is planted.

Step 4. Knowing life of orchard and years to full production from Step 1, calculate the break-even number of acres for each alternative. In this step, we calculate the break-even number of acres to be planted per year. In our example, the estimated number of acres to be planted annually changed frequently throughout the planning period. By knowing the number of break-even acres, however, management levels can be changed appropriately without additional calculation. For example, it might be that seed orchards and progeny testing require a planting program of 1000 acres per year to break even, while a seed orchard alone might require a planting program of only 500 acres per year. If the progeny tests had not yet been established, and the planting program fell from 1500 acres per year to 750, (and all the other assumptions remained consistent), it would be clear that, without any additional analysis, the progeny tests should not be established.

In this example, this step requires the following formula (devised by economist Kendrick Greer of the USDA Forest Service). Although we are only using one gain source in this example, this generalized formula can be used for more that one gain source. BEA is break-even acres, MVG/A_j is the per acre increase in value due to the genetic gain considered for gain source j (j=1,2,.., k), t_j is the first year that plantations are planted with the stock from gain source j, and n_j is the total number of years that stock from gain source j will be planted.

$$BEA = \frac{PV\ Costs}{\sum_{j=1}^{k} \left(\frac{MVG/A_j \left[1 + \frac{(1+i)^{n_j-1} - 1}{i(1+i)^{n_j-1}} \right]}{(1+i)^{t_j}} \right)} \qquad \text{Eq. 8.2}$$

Box 8.1 Time value of money: seedling netting (Vexar) maintenance

Example a. By planting seedlings collected from superior stands identified from progeny tests (and by using defined nursery and site prep practices), we expect that we can have final netting removal at age 3, and not need to incur the costs of vexar maintenance (assumed to be $22 per acre) for years 4 and 5. What is the present value per acre of these savings (present value of the benefits, PVB)?

$$PVB=\sum_{t=0}^{n}\frac{B_t}{(1+r)^t} \qquad \text{Eq. 8.3}$$

$$PVB(vexar\ savings)=\frac{\$22}{(1.04)^4}+\frac{\$22}{(1.04)^5}=\$36.89$$

b. The first commercial entry in the same stand is at age 50. What increased value at that time would equal the savings from two fewer years of netting maintenance? (both discounted to time 0).

$$\$36.89=\$\frac{X}{(1.04)^{50}}$$

$$X=\$262.16$$

If the interest rate were .06 instead of .04, the PVB of vexar savings would be $33.86 and the equivalent needed at age 50 to obtain the same present value would be $679.52. This example illustrates the fact that a small difference or savings at the beginning of the rotation can be worth more than large increases in value or savings at 50 years or beyond. In the Bovine Breeding Zone example, the PVB of $33 is greater than the MVG/A for all except the mixed conifer tractor category at the 4% interest rate.

Cont'd next page

Box 8.1 cont'd

In terms of relative risk, the savings from plantation establishment occurring during the first five years, must be less risky than increases in value of commercial timber obtained at 40-100 years. The 4% interest rate may not accurate reflect the relative risks of increases in value versus decreases in early costs.

"Accumulated establishment costs far exceed management cost in a plantation operation, making establishment costs one of the prime targets for cost reduction." Carlisle & Teich (1978).

In Eq. 8.2, the sum of annual discounted values factor is preceded by a "+1" and uses "n-1" as the power. This contrasts with the standard formula for the equation that does not use the "+1" and uses "n" as the power. The reason for this is that in the above formula, the sum of annual discounted values factor is being applied to a value that is already a present value; by definition, then, the value occurs at the beginning of each time period. The standard factor treats each value as if it occurs at the end of each period (C. Hansen, personal communication)[3].

The present value of the costs of the seed orchard and progeny tests is assumed to be $300,000 (from step 2), the interest rate is 4%, the seed orchard lasts 30 years and the first seedlings are planted in the field when the orchard is 16 years old. The MVG/A in this example

---

[3]*C. Hansen, Mount Baker-Snoqualmie National Forest, Seattle, WA.*

is $42 per acre, and the break-even point is planting 716 acres per year.

$$\textit{Break-even acres/year} = \frac{\$300{,}000}{\left[\dfrac{\$42 \times \left[1 + \dfrac{2.12}{.12}\right]}{1.87}\right]}$$

$$= \frac{\$300{,}000}{\$419.25} = 716 \textit{ acres/year}$$

The value $419.25 (the denominator in the break-even formula) is the incremental benefit of planting one acre per year over the productive life of the orchard (K. Connaughton, personal communication)[4].

In step 1, the 5% increase in height growth from making collections from open-pollinated parents in the field was shown in Fig. 8.2. To add this 5% gain to this example, it would be necessary to rerun the growth model. Assume that we had done this and the MVG/A for the 5% increase was $16, n is 7 and t is 9 years. Then the value $70.17 is obtained from the denominator of the break-even formula. This is the incremental benefit of planting one acre per year through the seven years seedlings would be planted from open-pollinated parents. The increment of $70.17 can then be added to the $419.25 from the seed

---

[4]*Research Economist, Pacific Northwest Research Station, Portland, Oregon.*

orchard for a total of $489.42 for the total program. Then the PVC, $300,000 is divided by $489.42 for a break-even of 613 acres/year. As many other benefits as desired can be added to obtain new break-even figures in this manner.

Step 5. Multiply PVB times the actual acres/year planted to get NPV and B/C ratios. The incremental benefits of planting one acre per year (step 4) can be multiplied by the number of acres planted per year to get the present value of the benefits for a project. This can then be compared with the present value of the costs for calculating the net present value or the benefit cost ratio. This simplified calculation will be accurate only if 1) the rotations are the same, or 2) only one rotation at a time is considered. As stated above, to take full advantage of decreasing the rotation through several rotations, a multiple rotation approach should be used.

The incremental benefit of planting one acre per year is $489.42 (step 4). If the zone actually had 300 acres per year planted, then the PV of the benefits would be 300 x $489.42, or $146,826. If the PV of the costs were $300,000, then the NPV of the seed orchard and progeny test option would be -$300,000 + $146,826= $-153,174. The benefit-cost ratio would be $146,826/$300,000 or .49. If the zone had 800 acres per year planted, the PV of the benefits would be $489.42 x 800 or $391,536. If the PV of the costs were still $300,000, then the NPV would be $391,536-$300,000= $91,536. The B/C ratio would be $391,536/$300,000 or 1.31. As described above in the break-even analysis (step 4), if the acres planted were 613, the NPV would be 0, and the B/C ratio would be 1.

Box 8.2 Evaluating the use of polycross seed

Should a manager use a polycross or open-pollinate in a mass selection plantation for reforestation seed? Alternatives: one polycross will yield 5 years worth of seed; one polycross will yield 10 years worth of seed.

Assume mean height is 50 feet, σ_p is 12.5 feet, h_i^2 is .20. The selection intensity for open pollinated seed is 2.2, so that using the standard gain formula (van Buijtenen, this volume), gain is 2.2 x .20 x 12.5 = 5.5 feet. Due to contamination from outside pollen (assumed to be 50%) the estimate is adjusted by multiplying by .75 to yield 4.1. Dividing this gain by the mean, 50, gives a percentage gain from open pollination of 8%.

For the polycross, the selection intensity is assumed to be 2.9, so that 2.9 x .20 x 12.5 is 7.25 feet, divided by the mean 50, is 14.5%.

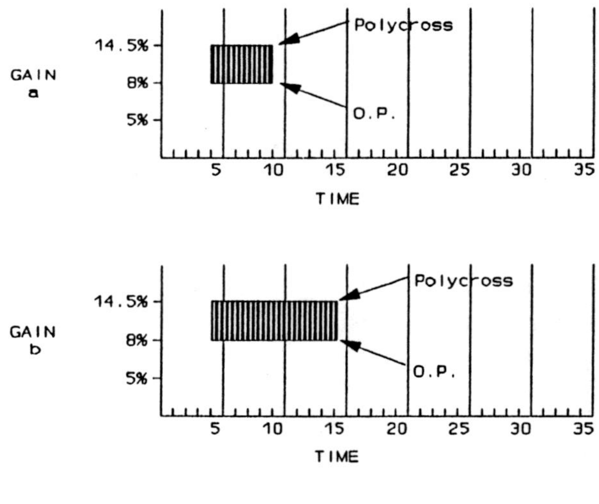

Fig 8.3. Gain/time diagrams for polycross compared to open-pollination in a mass selection plantation. The shading indicates the increase in gain from using the polycross, a) if seed lasts five years, and b) if seed lasts 10 years.

Cont'd next page

Box 8.2 cont'd

The gain begins in year 5 because it is assumed that pollination occurs in year 1, seed is collected in year 2, it is sown in year 3 and planted in year 5. This assumes a 2 year cone cycle, and two years in nursery (2-0) before outplanting. The marginal value per acre from the polycross is the value of 14.5% gain in height minus the 8% gain from open-pollinated, or 6.5% increase in height. By rerunning growth models at 14.5% and 8% about the base level, we arrived at an MVG/A of $20 per acre. To calculate the PVB per acre for the difference requires two steps for discounting, one to discount the five years to the first year of the five (the numerator), and then to discount from the beginning of the benefits to the present, when the polycross would have to be done (the denominator).

$$PVB/acre = \frac{\$20 \ x \left[1 + \frac{(1.04)^4 - 1}{.04(1.04)^4}\right]}{(1.04)^5}$$

$$= \frac{\$20(4.63)}{1.22} = \$75.90$$

$$\$75.90 \ x \ 500 \ acres/year = \$37{,}950$$

If the seed lasts 5 years, you could afford to spend up to $37,950 discounted to year 0 to make polycrosses.

Cont'd next page

Box 8.2 cont'd

If the seed lasts 10 years, all that changes is the exponent on 1.04 in the numerator:

$$PVB/acre = \frac{\$20 \; x \left[1 + \frac{(1.04)^9 - 1}{.04(1.04)^9}\right]}{(1.04)^5}$$

$$= \frac{\$20(8.44)}{1.22} = \$138$$

$$\$138 \; x \; 500 \; acres/year = \$69{,}000$$

If the seed from one polycross lasts 10 years, you could afford to spend up to $69,000.

Step 6. Conduct sensitivity analyses on appropriate parameters. Sensitivity analyses should be conducted on parameters for which there is uncertainty in their estimation. This is done by changing the values and running the analyses again. Table 8.2 shows a simple example of a sensitivity analysis using some of the data from above, but with different present values of costs. In Table 8.2, break-even acres were calculated for varying MVG/A, interest rates (which affect MVG/A), 0 and 1% real price increase, present values of costs, years to first planting of orchard seedlings, and total years that seedlings are planted from orchard seed. At the standard 4% interest rate, break-even acres are relatively sensitive to the 1% real increase in stumpage. The change in MVG/A is from $42 to $69 per acre, which decreases

Table 8.2. Planting acres per year needed to break even under different assumptions in the Bovine Breeding Zone.

Interest Rate	Real Price Increase[1]	MVG/A	First Seedlings[2]	Total Years[3]	Present Value Costs[4]	Break-even Acres/Yr
4%	1%	$69	16	30	391,088	590
	1%	$69	14	30	391,088	546
	1%	$69	18	30	391,088	638
	1%	$69	16	50	391,088	475
	0	$42	16	30	391,088	970
	0	$84	16	30	391,088	485
	0	$42	16	30	195,554	485
6%	1%	$25	16	30	365,264	1001
	0	$16	16	30	365,264	1565
	0	$32	16	30	365,264	782
	0	$16	16	30	182,632	782
8%	1%	$10	16	30	347,048	2854
	0	$ 6	16	30	347,048	4757
	0	$ 6	16	30	173,524	2379
	0	$12	16	30	347,048	2379

[1] 1%/year real stumpage value increase for 50 years.

[2] First year that seedlings from the orchard are planted.

[3] Total number of years that seedlings are planted from orchard seed.

[4] Cost estimates for 11 acre seed orchard and 72 acres of progeny tests (Vicky Erickson, Umatilla N.F.). Estimates may be high.

the break-even acres from 970 acres per year to 590. On the other hand, under a 1% price increase, break-even acres per year are less sensitive to two year changes in the time to the first planting with seed orchard stock (546 if 14 years, 590 if 16 years, 638 if 18 years). This approximates a 50 acre difference for each two year time period. Extending the time the seed orchard produces from 30 to 50 years similarly decreases the break-even acres from 590 to 475, about 115 acres/year. Since the present value of the costs is directly related to the break-even acres, halving the costs leads to reducing by one half the break-even acres. For example, with constant prices (0% price increase), break-even acres would be 970 if the present value of the costs were $391,088, but only 485 if the costs were one half that. GRM options which lower costs can have a significant effect on break-even acres, at least in this example.

8.3 Choices of GRM Alternatives under Constraints

Many GRM alternatives can be proven to be cost-effective using the techniques described above. Yet budget or other factors may limit the amount of work that can be done.

8.3.1 Limited by one resource

Buongiorno and Gilless (1987, pgs. 239-240) suggest a simple method for choosing projects limited by one resource. Let R_m be the total amount of dollars available in a given year, with R_i the amount of that resource needed for project i (i=1,... n, where n is the number of projects) in a given year. They define the net present value-resource ratio for project i, NPR_i as:

$$NPR_i = \frac{NPV_i}{R_i}$$

Let's assume that funding is limited and that four different breeding zones are competing for funding for progeny test, seed orchard, or mass selection plantation establishment in one year. Let breeding zone 1 have a progeny test to grow in a greenhouse, which will cost \$10,000 the first year, and the NPV is \$2,000,000. Breeding zone 2 has progeny test material to grow bare-root, and the cost for the first year is \$15,000, with an NPV of \$1,000,000. Breeding zone 3 needs \$3000 to establish stock for mass selection plantations, with an NPV of \$500,000. Breeding zone 4 needs a site selected for a seed orchard, with site selection cost \$5000, for an expected NPV of \$1,000,000. Hence, $NPR_1=200$, $NPR_2=67$, $NPR_3=167$, and $NPR_4=200$.

Buongiorno and Gilless then suggest ranking all projects in decreasing order of their net present value-resource ratio, and going down the list until the resource is exhausted. For example, if there was a total available of \$20,000 for new starts, 4 and 1 would rank equally, so that 4 and 1 and 3 could be done with \$3000 left over. If only \$10,000 were available, the choice would have to be between doing 4 with some dollars remaining, or doing only 1. The dollars remaining reflect fact that these types of projects are not divisible.

One of the difficulties with this technique is that, heuristically, for breeding zone 1, once the \$10,000 is in, there is a commitment to planting, site selection, and all the other expenses that might occur in different years, which will require funding to be available in larger quantities in years to come. For breeding zone 4, that commitment is not yet made; but on the other hand, all the expenses for seedlings remain to be spent. Thus, ranking by benefit/cost ratios or considering both the benefit/cost and the net present value/resource ratio might be appropriate.

8.3.2 Limited by more than one resource: mathematical programming
The goal of these techniques is to find the set of projects that maximizes total net present value without exceeding the resources available (Buongiorno and Gilless 1987). Complete descriptions of linear programming techniques in forestry can be found in this book, along with useful annotated references for examples at the end of each chapter. In this chapter, only applications for GRM programs will be reviewed.

The first effort in linear programming in GRM programs was that of van Buijtenen and Saitta (1972). They used this technique to examine different investments in seed orchard management and their profitability. A more holistic approach was used by van Buijtenen *et al.* (1975), in which a linear programming model was developed to analyze the effects of modifying wood quality by genetic and silvicultural methods along with the effects of papermaking technology.

In contrast, Porterfield (1976), and Porterfield *et al.* (1975) used *goal programming* to evaluate an optimum combination of selection intensities for a group of traits. In these analyses, the author rejected the use of linear programming because it might be impossible to meet the stated improvement levels within the limit for selection expenditures. He felt that 10 percent improvement in specific gravity, for example, is a goal, rather than a constraint. A single objective function did not fully describe the situation; rather, profit, for example, is one of many possible goals. In goal programming, the deviations from stated goals are minimized. The tree improvement programs in the study could produce greater gain by doubling the selection expenditure, using higher roguing intensity, and selection of rust-free trees in infected areas. By using goal programming, Porterfield was able to bring some companies' selection practices more into line with their stated goals.

Chapter 12 in Buongiorno and Gilless (1987) describes dynamic programming using multistage decision-making. This seems like a useful approach in GRM programs. First, there are many nonlinear relationships in GRM programs, such as nonlinear increases in value with diameter, and a nonlinear relationship between selection intensity and cost. Thus, not requiring linearity may be an asset in developing realistic models. The multistage decision-making aspect is also important; as in the silviculture example given in Chapter 12 of Buongiorno and Gilless, decisions in GRM programs cannot be made independently of what has been done earlier, and what will be done later. By developing the model for different decision points, we can determine the flexibility of our current choices to changing conditions in the future, as we shall see in the next section.

8.4. Risk and Uncertainty in GRM Decisions

Decisions for tomorrow's forests must be made by managers today. Decisions are made under conditions of certainty, risk, or uncertainty (Marty 1964). Decisions are made under certainty if the result of each possible action is a single, known, outcome. For example, the decision to purchase U.S. Government savings bonds may be considered to fall into this category. If there is more than one possible outcome, but the objective probability of each outcome is known, the decision is made under conditions of risk. Finally, if the objective probabilities are unknown, then the decision is made under uncertainty.

One of the few certainties in GRM programs is that most decisions are made under conditions of risk or uncertainty. Lundgren (1983) described different ways of dealing with uncertainty. One can ignore it, or postpone making the decision until more information is available. Both these options have been, and continue to be used by GRM managers. Sensitivity analysis is a fairly common way of dealing with uncertainty in the genetics literature.

For example, Fins and Moore (1984) used sensitivity analysis to examine three different potential growth patterns for improved stock, where, compared to unimproved stock, the improved stock had 1) a constant volume advantage, 2) a constant proportional advantage, and 3) a maximum volume advantage of about 18 percent at age 20, and then a slow decrease in proportional advantage until volumes were approximately equal at maturity. More commonly, sensitivity analysis is used for stumpage, interest rates, costs, selection intensity, seedling production rate per unit of seed, seed yields, and years to seed production (Kurinobu 1987; Ledig and Porterfield 1981). In the above papers, factors influencing the cost-effectiveness of the current tree improvement programs are examined. In Cornelius and Morgenstern (1986), in addition to the above considerations, trade-offs between plus tree selection and family selection in both costs and benefits were also examined.

Another use of sensitivity analysis is to examine different alternative scenarios within GRM programs. Gill (1983) compared seed orchard and cutting alternatives using differing gain values, pessimistic, expected, and optimistic; cutting output low or high; orchard productivity pessimistic, expected and optimistic; with project lengths of 19, 29 or 39 years.

Yet alternative scenarios cannot be restricted to within the GRM program itself. Suppose that delays in program funding kept the first commercial thinning postponed for 10 years on 50% of the acreage? Suppose a wildfire burned 5% of the acres in a bad fire year? Suppose that subunits of the acreages suffer from insect and/or disease outbreaks? What if mixtures of species are planted or naturally regenerated so that only 75% of the harvested trees on an acre are from improved stock? What if projected "commercial" second growth thinnings turn out not to be commercial?

The following discussion on alternative approaches to risk and uncertainty is from an unpublished annotated reference list by James Cathcart of the Pacific Northwest Research Station (USDA Forest

Service) in Portland, Oregon. A complete review of the literature can be found in Cathcart (1989).

One important aspect of any risk analysis is whether the decision unit (i.e. person, firm, agency) is risk-neutral or risk-averse (Cathcart 1989). Risk neutrality is defined as indifference to risk (e.g., variability in possible return outcomes). If an investor is risk-neutral, only the mean outcome for each individual decision needs to be determined (i.e., average net present value); while if the decision-maker is risk-averse, the analysis should concentrate on estimating the entire probability distribution of investment outcomes (mean, variance, skewness). For example, a risk-neutral decision unit would prefer the alternative with the highest mean, while a risk-averse individual would likely prefer an alternative with a lower mean if it is accompanied by less risk (i.e., lower variance of return).

Imagining alternative scenarios for the outcome of future events, and examining the impacts on profitability is one way to handle uncertainty. Probabilities representing the likelihood of each scenario being true could also be estimated. Then a probability distribution of NPV's could be estimated based on the alternative's outcome under each scenario and the likelihood of that scenario being true. Sensitivity analysis could then consider a different set of probabilities for the likelihood of each scenario being true (Cathcart 1989).

Simulation can also be used, but may have drawbacks from the standpoint that models tend to be very detailed, which prohibits a large scale sensitivity analysis regarding the estimated probability distribution for each parameter. Chambers *et al.* (1986) use the triangular probability distribution, which provides a formal representation for the likelihood of the worst case, best case, and most likely case scenarios. In Cathcart's opinion (personal communication), this approach strikes a middle ground between the detail in simulation and the flexibility in scenario analysis.

Bentley and Kaiser (1967) and Talerico *et al.* (1978) used a decision-tree approach to risk. This may be particularly applicable to GRM program decisions. Decision-trees allow for much more flexibility in decision-making than does the conventional approach, which lays out all the activities as if all future decisions need to be made today. In tree improvement, for example, mass selection plantations or seed orchards may be established today, but decisions to rogue, not rogue, or do supplemental mass pollination need not be made until the trees begin to reproduce.

McQuilkin (1984) took a safety first approach to uncertainty, whereby the best decision minimizes the effects of the worst possible outcomes or minimizes the probability of the worst outcome. Thompson (1968) gave a good review of some heuristical approaches to risk and uncertainty, such as minimizing regret in the even that the wrong decision is made. This author's (Friedman's) personal favorite subjective criteria is the "least wasted effort" by which the alternative that leads to the least wasted effort under a broad range of scenarios, and meets the minimum criteria for improvement and/or stewardship goals, is chosen.

Finally, from Field (1988) a slightly different view:

"I have often argued that one of the most promising ways to deal with the uncertainty of the biological world but also the marketplace is to slant forest management decisions in favor of quality and diversity... Forest managers can approach the fundamental problem of long-term risk through diversification of the products produced from forest lands. We may speak of woodlands as portfolios, and diversify within a woodland by species and products of timber and other commodities. We can also speak of portfolios of woodlands, where one may specialize within an individual woodland but diversify among them. Or, the manager of a large forest holding may manage a portfolio of timber compartments and diversify between and among them."

When the reader becomes comfortable with the techniques presented in section 8.2, to the point of recognizing their limitations in dealing with the uncertain world (and geneticists' uncertain estimates of the uncertain world) application of the concepts reviewed here would be the next step.

Acknowledgements

The author expresses thanks to the economists Kendrick Greer and Kent Connaughton for help with earlier versions of this paper. In addition, the patient, encouraging and thorough reviews of Jim Cathcart and Chris Hansen are deeply appreciated. Thanks to Don Wood of the Ochoco National Forest for the growth model runs for the original Bovine analysis. To fellow geneticists Carol Aubry and Vicky Erickson thanks for continuing ideas, information, support and enthusiasm.

Literature Cited

Bentley, W.R.; Kaiser, H.F. 1967. Sequential decisions in timber management — a Christmas tree case study. *J.For.* 65(10):714-719.

Bridgwater, F.E.; Squillace, A.E. 1986. Selection indexes for forest trees. In *Proc. Southern Coop. Series Bull. #309,* June 1986, Louisiana Agricultural Exp. Sta., pp 19-20.

Bridgwater, F.E.; Stonecypher, R.W. 1979. Index selection for volume and straightness in a loblolly pine population. In *Proc. 15th S. For. Tree Improv. Conf.* Mississippi State Univ., Starkville, Miss. pp 132-139.

Buijtenen, J.P. van; Saitta, W.W. 1972. Linear programming applied to the economic analysis of forest tree improvement. *J. For.* 70(3):164-167.

Buijtenen, J.P. van; Alexander, S.D.; Einspahr, D.W.; Ferrie, A.E.; Hart, T.; Kellogg, R.M.; Porterfield, R.L.; Zobel, B.J. 1975. How will tree improvement and intensive forestry affect pulp manufacture? *Tappi* 58(9):129-134.

Buongiorno, J.; Gilless, J.K. 1987. *Forest Management and Economics.* MacMillan Pub. Co., New York, NY. 285 pp.

Carlisle, A.; Teich, A.H. 1978. Analysing benefits and costs of tree-breeding programmes. *Unasylva* 30:34-37.

Cathcart, J.F. 1989. Evaluating risk-adjusted discount rates in forest investment evaluation. Ph.D. Thesis. Virginia Polytechnic and State University, Blacksburg, Virginia. 280 p.

Chambers, P.C.; S.A. Sinclair; C.C. Hassler; B.G. Hansen. 1986. Forest products investment model: a microcomputer tool for incorporating risk into capital budgeting. *Forest Products Journal* 36(1):64-68.

Connaughton, K.P. 1986. Economic considerations for forest land management: one economist's view of U.S. public ownership. In Oliver, C.D.; Hanley, D.P.; Johnson, J.A., eds. *Douglas-fir: Stand Management for the Future:* Proceedings of a symposium; 1985 June 18-20; Seattle, WA, pp.46-50.Contribution no. 55. Seattle: College of Forest Resources, University of Washington.

Cornelius, J.P.; Morgenstern, E.K. 1986. An economic analysis of black spruce breeding in New Brunswick. *Can. J. For. Res.* 16:476-483.

Clutter, J.L.; Fortson, J.C.; Pienaar, L.V.; Brister, G.H.; Bailey, R.L. 1983. *Timber Management: A Quantitative Approach.* John Wiley & Sons, New York.

Field, D.B. 1988. An investment analysis of Maine forest lands. Presented at the annual meeting of Pingree Associates, June 23. University of Maine (unpublished), 12 p. Department of Forestry, Orono, Maine.

Fins, L.; Moore, J.A. 1984. Economic analysis of a tree improvement program for western larch. *J. For.* 82:675-679.

Ford, D.; Merrill, R.E.; Mohn, C.A.; Stine, R.A. 1983. Economic analysis in planning the Minnesota Department of Natural Resources tree improvement program. In *Proc. North Central Tree Improv. Conf.*, Madison, Wisc., pp 128-136.

Gill, J.G.S. 1983. Comparisons of production costs and genetic benefits of transplants and rooted cuttings of Picea sitchensis. *Forestry* 56(1):61-73.

Hansen, C. 1990. A model for economic analysis of genetic tree improvement programs. USDA Forest Service, Mt. Baker-Snoqualmie National Forest. (In preparation).

Kurinobu, S. 1987. Cost and benefit analysis of first-generation tree improvement programs for hinoki (Chamaecyparis obtusa) in Kyushu. *J. Japanese For. Soc.* 69(11):435-444.

Ledig, F.T.; Porterfield, R.L. 1981. West coast tree improvement programs: a break-even, cost-benefit analysis. USDA Forest Service Res. Pap. PSW-156. 8 p. Pacific Southwest Experiment Station, Berkeley, Calif.

Lundgren, A.L. 1983. Strategies for coping with uncertainty in forest resource planning, management, and use. In *Proc. Society of American Foresters Conv.*, Portland, Ore., pp. 574-578.

Marty, R. 1964. Analyzing uncertain timber investments. USDA Forest Service Research Paper NE-23, 21 p. Northeastern Forest Experiment Station, Upper Darby, Pennsylvania.

McDivitt, J.; Hansen, C. 1983. *Proceedings of the National Silviculture Workshop, Economics of Silvicultural Investments.* U.S.D.A. Forest Service Timber Management, Washington, D.C.

McKenney, D.; van Vuuren, W.; Fox, G. 1988. The economics of tree improvement: a comparison of clonal forestry and seed orchards for black spruce in Ontario. Report, Department of Agriculture Economics and Business, University of Guelph, Guelph, Ontario.

McQuilkin, R.A. 1984. Pruning guide for black and white oaks. *J. For.* 82(8):487-489.

Paques, L.E. 1984. Tree improvement strategies: modelization and optimization. M.S. Thesis. N.C. State Univ. Raleigh, NC.

Porterfield, R.L. 1975. Economic aspects of tree improvement programs. In *Proc. 24th Forestry Symp.,* Louisiana State Univ., Baton Rouge, LA, B.A. Thielges, ed. Forest tree improvement- the third decade. Part 2. Applying genetic gains to forest management, pp. 99-117.

Porterfield, R.L. 1976. A goal programming model to guide and evaluate tree improvement programs. *For. Sci.* 22(4):417-430.

Rehfeldt, G.E. 1989. Expecting the predicted. Abstract from Western Forest Genetics Association Meeting, 1989, Clarkston, Washington.

Talerico, R.L.; Newton, C.M.; Valentine, H.T. 1978. Pest-control decisions by decision-tree analysis. *J. For.* 76(1):16-19.

Thompson, E.F. 1968. The theory of decision under uncertainty and possible applications in forest management. *For. Sci.* 14(2):156-163.

Suggestions for Further Reading

Bergman, A. 1970. Evaluation of costs and benefits of tree improvement programs. *Unasylva* 24(2/3):89-95.

Brett, P.; Benskin, H. 1978. Guidelines for cost-effective intensive forest management in B.C. In *Proc. 59th Annual Meeting of the Woodland Section,* Canadian Pulp and Paper Assoc., Montreal, QUE., pp. 1-6.

Caraher, D.L. 1986. One-page risk analysis: a student's index for evaluating risk in simple management and project situations. Unpublished report, Management Consulting Group, USDA Forest Service, Portland, Oregon.

Carlisle, A.; Teich, A.H. 1970. The cost benefits of tree improvement programs. Can. For. Serv. Rep. PS-X-20, Chalk River, ONT. 28 p.

Carlisle, A.; Teich, A.H. 1976. The economics of tree improvement. In *Proc. 15th Canadian Tree Improvement Assoc. Conf.* Ottawa, Canada, pp. 42-56.

Cotterill, P.P. 1986. Genetic gains expected from alternative breeding strategies including simple low cost options. *Silvae Genet.* 35:212-223.

Dutrow, G.F. 1974. Economic analysis of tree improvement: a status report. USDA For. Serv. Gen. Tech. Rep. SO-6. 10 p.

Fight, R.D.; Snellgrove, T.A.; Curtis, Robert, O.C.; DeBell, D.S. 1986. Bringing timber quality considerations into forest management decisions: a conceptual approach. In Oliver, C.D.; Hanley, D.P.; Johnson, J.A., eds. *Douglas-fir: Stand Management for the Future: Proceeding of a Symposium;* 1985 June 18-20; Seattle, WA, pp. 22-25. Contribution no. 55. Seattle: College of Forest Resources, University of Washington.

Hasnain, S.; Cheliak, W. 1986. Tissue culture in forestry: economic and genetic potential. *For. Chron.* 62(4):219-225.

Kellogg, R.M. 1982. Coming to grips with wood quality. *For. Chron.* 58(6):254-257.

Layton, P.A. 1978. Comparison of two advanced generation mating designs for loblolly pine. M.S. Thesis. Texas A&M Univ., College Station, Texas.

Namkoong, G.; Johnson, J.A. 1986. Influence of the value function on genotype-by-environment relations. *Silvae Genet.* 36(2):92-94.

Porterfield, R.L.; Ledig, F.T. 1977. The economics of tree improvement programs in the northeast. In *Proc. 25th Northeastern Forest Tree Improv. Conf.*, Orono, Maine, pp. 35-47.

Porterfield, R.L.; Zobel, B.J.; Ledig, F.T. 1975. Evaluating the efficiency of tree improvement programs. *Silvae Genet.* 24:2-3.

Reilly, J.J.; Nikles, D.G. 1977. Analysing benefits and costs of tree improvement: Pinus caribaea. In *Proc. Third World Consultation on Forest Tree Breeding,* Canberra, Australia, pp. 1099-1124.

Riemenschneider, D.E. 1979. The genetic and economic effect of preliminary culling in the seedling orchard. In *Proc. 13th Lake States Forest Tree Improv. Conf.*, pp. 81-91.

South, D.B. 1986. Economics of seed efficiency. *J. For.* 84:33-34.

South, D.B. 1987. Economic aspects of nursery seed efficiency. *So. Jour. Appl. For.* 11(2):106-109.

Sutton, W.R.J. 1984. New Zealand experience with radiata pine. The H. R. MacMillan Lectureship in Forestry. University of British Columbia, Vancouver. 21 p. Talbert, J.T.; Weir, R.J.; Arnold, R.D. 1985. Costs and benefits of a mature first-generation loblolly pine tree improvement program. *J. For.* 83(3):162-165.

Thomson, T.A., Lester, D.T., Martin, J.A. and Foster, G.S. 1989. Using economic and decision making concepts to evaluate and design a corporate tree improvement program. *Silvae Genet.* 38(1):21-27.

Williams, C.; de Steiguer, E. 1990. Choosing a strategy for advanced generation seed orchards. *For. Sci.* 36(1):156-168.

Chapter 9
Developing Seed Transfer Zones

R. D. Westfall[1]

To manage breeding programs, we must optimize adaptations to economic constraints and biological conditions. Although a single variety would ideally be well adapted to all environments, such a condition rarely exists for breeding programs in most species. Important genotype x environment (G x E) interactions usually exist, resulting from changes in relative performance rank among genotypes, so that genotypes and environments must be specified separately (Gregorius and Namkoong 1986). As a consequence, seed sources are usually subdivided into zones within which rank changes are minimal. Even when there are no changes in ranking in individual traits, multiple trait indices can result in rank changes among sources (Namkoong 1985, Namkoong and Johnson 1987), thus generating the need for many breeding zones.

A concept emphasized by van Buijtenen in Chapter 2 is worth reiterating: the area where seed sources (provenances) are collected and the area where propagules from these sources are distributed can be managerially and geographically distinct. The former area is

[1]*Biological Statistician, National Forest Genetic Electrophoresis Laboratory, USDA Forest Service, Placerville Nursery, Camino, CA*

L. Fins et al. (eds.), Handbook of Quantitative Forest Genetics, 313–398.

considered the seed zone and the latter the planting zone. But van Buijtenen introduced the following terminology: procurement and deployment zones, instead of seed and planting, respectively. The area defining these two zones could be identical or the procurement zone could be a subset of the deployment zone or vice versa. Alternatively, the two areas could be nonoverlapping, as is the case with deployment of exotic species.

9.1 Chapter Organization

Establishment of seed zones will be covered first in section 9.2, where I present testing methods designed to define genotypic patterns and analytical methods used to effectively describe patterns and provide information necessary to develop transfer guidelines. The section concludes with two approaches designed to set transfer limits, given the data from the previous step.

Once a breeding program has advanced into the first generation, the procurement zone becomes unnecessary, except as it applies to genetic conservation goals and additions to the breeding lines. Instead, attention is given to the relative performance of genotypic entries, such as families, over sets of environments within, and sometimes among, deployment zones. In the next section, 9.3, I discuss methods for describing genotypic stabilities in performance over environments, starting first with linear models and then providing some thoughts about applying nonlinear models. In the last section (9.4), I suggest experimental designs and analytical approaches that would help us more effectively deploy genotypes over environments.

9.2 Development of Seed (Procurement) Zones

Establishing seed zones is first a matter of examining patterns of genetic differentiation among provenance samples in common gardens. High differentiation suggests the need for multiple seed zones. Also important are changes in genetic patterns over many site environments. General approaches used to examine patterns include short- and long-term genetic tests. Under the former, the study period, although short, is assumed to be sufficiently long and the environmental conditions of the test sufficiently different to provide data appropriate for developing transfer rules. The latter assumes that the sites are random samples of commercially important environments, that important changes in seed source ranking will occur over planting site environments, and that these changes will be expressed only when the plantation is older.

To effectively describe patterns among seed sources, choice of analytical methods is important. I will present a number of approaches, from the visually descriptive to the statistical and from the univariate to their multivariate extensions. Because patterns in most traits tend to be correlated and variation in these traits tends to reduce to a few major patterns, I favor the multivariate techniques (for further arguments in support of this viewpoint, see Namkoong 1982).

Whatever analytical method is preferred, the importance of tabular and graphical examination cannot be over emphasized. With multivariate techniques in particular, it is all too easy to distance oneself from the data. But with the current generation of software for personal computers, output from statistical analyses can now be easily transferred to software for tabular or graphical presentation.

9.2.1 Experimental methods

9.2.1.1 Short-term tests

Under assumptions applied to short-term tests, seed sources are grown at one or two sites (either nursery or field or both) or under contrasting treatments (Rehfeldt 1986; Campbell 1986). Several traits, such as growth, phenology and cold hardiness, are measured. The population sampling method normally used entails sampling a few populations within the geographic range studied and collecting wind-pollinated seed from five to ten parents within each population. Alternatively, Campbell recommends sampling more locations in a grid or transect to minimize correlations among geographic variables. Under Campbell's recommendations, at about 60% of the locations, two parent trees are sampled in order to estimate additive variance within location and to test lack-of-fit in regression models. At the remaining locations, only one parent is sampled.

One can object to the approach of sampling a limited number of families per location in that additive variances may be heterogeneous. However, most studies do not include tests for homogeneity of variances. Also, the number of parents and progeny per population in most studies is smaller than the number needed to statistically differentiate additive variances among populations. There can be clinal differences in within-population variances, but Campbell (1986) addressed this possibility by testing population differences among variances within nursery plots, which contain the major portion of additive variance. This is done by computing within-plot standard deviations and testing these as differences among seed sources in the analysis of variance (ANOVA). If differences are significant, the standard deviation is included as a trait in further analyses.

Measures of source location (eg. latitude, elevation, and aspect) and their more complex terms (such as squared and crossed terms), are then regressed against traits measured. Campbell then computes the additive genetic correlation matrix among sources, extracts principal components (PC's) of this matrix (discussed later in section 9.2.2), and then regresses linear, quadratic, and cubic terms of location against significant PC's.

Short-term tests confer a number of advantages. First, a large amount of information can be gathered in a short time. Thus if correlations between these early test results and those of rotation age are high enough and are appropriate for use, the short-term tests can be used effectively for early selection. Secondly, correlations between a trait and the environment of origin imply adaptive variation, although biological explanations of patterns usually seem to be *post hoc*. Also, these results indicate the amount of variation and changes in traits over a wide geographic area and over environmental gradients. For example, differences among populations (sources) of Douglas-fir were greater over east-west gradients than north-south (Campbell 1986).

There are limitations to the use of short-term tests. First, response functions describing source performance might not remain the same over all planting sites. For example, 29-year patterns in an elevational transect in ponderosa pine (taken from Fig 1, Conkle 1973) suggest that elevational differences were small at the high-elevation site, but were steeper and opposite in direction at both the low- and mid-elevation sites (Fig 9.1)[2].

Thus if we were to select sources within the top five feet of the maximum, the optimum sources would fall within a range of 1500 ft of elevation for the low elevation site, 2500 ft at the mid elevation site, and the entire range of sources at the high elevation site. To use a single criterion at all sites would be inappropriate; differentiation is high at the low elevation site and very low at the high elevation site. However, the point here is that transfer risks vary by site location along the elevational gradient. Furthermore, there is substantial genotype by environment interaction and therefore, a single site does not indicate optimum sources for the other sites.

Another complication is that of age-related stability of geographic patterns. Namkoong et al. (1972) showed that early patterns among provenances might not be well-correlated with late patterns. In this instance with Douglas-fir, the transition in patterns from early to late occurred around the fifteenth year.

---

[2] *Data are fitted with fourth-order equations, so that asymptotic regions would be indicated. Thus there is some overfitting.*

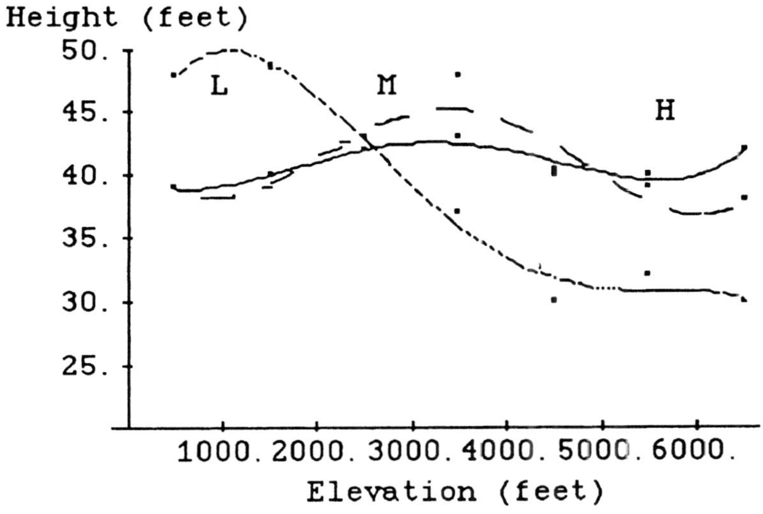

Figure 9.1. Fitted curves for seed source height in ponderosa pine at low (L, 960 ft), middle (M, 2730 ft), and high (H, 5650 ft) elevation sites (from Conkle 1973).

9.2.1.2 Long-term tests

Long-term tests are best exemplified by the range-wide provenance tests in many species, described by Wright (1978) and the south-wide seed source study of pine species in the southeastern USA of Wells and Wakely (there are many reports; see Dorman (1976) for a summary). Sources were equally represented in widely distributed, randomized, complete block experiments. Assuming sites that meet breeding objectives are used, such tests can describe patterns in genotype x environment interaction, define trait optima, capture information from such unanticipated or changing environments as disease or insect infestations, and estimate juvenile-mature correlations.

The drawbacks of long-term tests are the extended time required to recover information, the difficulty in collecting data on large trees, and the size and complexity of designs. As a consequence of the latter, tests are frequently limited to a few (2-5) sites, making extrapolation to the many commercial sites difficult.

9.2.1.3 Sampling methods and experimental design

With the exception of work done over the years in the Southern and Gulf States (see Van Buijtenen, Ch 2), there is need in the U.S. for more detailed studies, covering the full range of planting sites for the breeding program in each species. Because the objective of zone formation is to estimate the form of geographic patterns, I favor gridded seed source (provenance) sampling, as described above in Section 9.2.1.1. The approach is taken from response surface methodologies (Box and Draper 1987; Box et al. 1978), which would ensure even, unbiased distribution of sources. There should be limited numbers of duplicate samples at some locations so that adequacy of models and limits of transfer can be determined (see Section 9.2.3, below). If existing single-tree selections in a breeding program are evenly distributed over the geographic range of interest, seed from each of these could contribute to the study either *in toto*, by random draw, or be selected to approximate a grid.

The usual hierarchical sampling structure of more limited numbers of sources and more families per source is better suited for selection among the included sources, whereby sources are regarded as fixed rather than random as in the gridded structure. Here the focus is on adequate numbers of progeny to detect differences among source means. Generally, there should be 100 to 200 progeny in the test, distributed among about ten families per seed source (Cotterill 1990).

The size and scope of future tests should be guided by information from previous short-term and existing long-term tests. For example, if existing data suggest that genetic differentiation is low, studies might focus on a few, widely spaced sources, and on widely spaced sites. Test size per site can be reduced by using partially balanced incomplete blocks or overlapping unbalanced planting designs. An example of the latter (from Namkoong et al. 1988) is shown below.

	Site				
Source	1	2	3	4	5
1	x				
2	x	x			
3	x	x	x		
4	x	x	x	x	
5		x	x	x	x
6			x	x	x
7				x	x
8					x

Say for example, that both the sites and sources above are arrayed north to south. Then, for each site to the south, the northernmost source in the series is dropped and the next most southern source added. In the past, analyses of these designs were possible only with a mainframe computer. But, inexpensive, powerful statistical programs are now available for mini- or microcomputers, making analysis of such highly unbalanced designs more feasible.

The hierarchical and grid sampling structures advocated above create a more complicated experimental design than those described in Chapters 4 and 6. A full analysis of variance and expectations of means squares for s sources and f families in sources replicated in b blocks at a single location is shown in Table 9.1.

If more than one site is tested, additional sources of variation would be Site, Site by Source, and Site by Family(Source). Note also that blocks would be nested within sites. From the expected mean squares (Table 9.1), there is no pure error term for sources, either in the single or the multi-site design. To test sources, a *quasi* F statistic must be constructed along with approximate degrees of freedom (Steele and Torrie 1980:357). An appropriate F ratio for sources in Table 9.1 is:

$$F = \frac{MS_S + MS_E}{MS_{BS} + MS_{F(S)}}$$

where MS_S, MS_E, MS_{BS}, and $MS_{F(S)}$ are the mean squares for sources, error, block by source, and families in sources, respectively. In a sin-

Table 9.1. Analysis of variance of plot means for a mixed model of seed sources and families within sources, replicated in blocks. Coefficients for variances assume a balanced design.

Source of variation	df	Expected mean squares
Block	b - 1	$\sigma^2_e + f\sigma^2_{bs} + fs\sigma^2_b$
Source	s - 1	$\sigma^2_e + f\sigma^2_{bs} + b\sigma^2_{f(s)} + bf\sigma^2_e$
Block by source	(b - 1)(s - 1)	$\sigma^2_e + f\sigma^2_{bs}$
Family (source)	s(F - 1)	$\sigma^2_e + b\sigma^2_{f(s)}$
Plot error	s(b - 1)(f - 1)	σ^2_e

Where:

b = number of blocks

s = number of seed sources

f = number of families per source

σ^2_e = $\sigma^2_w/n + \sigma^2_{bf(s)}$

σ^2_w = variance among trees within plots

n = number of trees per plot (or the weighted mean if the number of trees per plot are unequal).

gle site experiment where errors are controlled, the block by source interaction sums of squares can be pooled with the plot error, simplifying analysis. But to pool the site-related interaction terms with the plot error in a multi-site experiment would be inappropriate if such interaction estimates were significantly greater than zero.

Families allocated randomly within blocks would result in a fully randomized design, where trees from differing sources would compete. When the goal of the experiment is to minimize competition among sources, families should be planted in source plots, whereby sources are randomly allocated to plots in each block and then families randomly allocated within source plots. This method of plot allocation results in a split-plot design. Although analysis of the design and estimation of variance components would not be altered, there would be some reduced precision in the test for differences among sources.

9.2.2 Analytical approaches

If geographic patterns are simple and changes in rank are negligible, it suffices to compare the ranks of provenance means across sites. Even when patterns are more complex, the highest-ranking sources over locations or groups of locations can be used to make rough guidelines for seed zones. For example, Westfall et al. (1977) compared the top ten eastern white pine provenances among sites in the northeastern USA. Areas of the region that uniformly included the same set of provenances embodied a zone. An extension would be to compute correlations between pairs of sites (Burdon 1977), with sites as the variables and source means at each site as the observations. Sites where provenance means are well correlated would compose a breeding zone. The size of the correlation necessary to form a zone depends on subjective criteria. I favor a confidence interval for the correlation coefficient that includes 0.9 and above (Steele and Torrie 1980:278-279); those below 0.9 suggest that important rank changes may exist. This and the preceding method are similar in principle to, but less sophisticated than the cluster analysis of planting locations (Abou-El-Fittouh et al. 1969). These correlation matrices also can be used to construct best linear predictor (BLP) indices for source selection (Harvey and Townsend 1985; also see Ch 6).

Errors from many sources, experimental or biological, often obscure patterns that exist in the data. Furthermore, measured traits are normally correlated with one another and these correlations sometimes obscure patterns in the data (this is the justification for using the analysis of covariance). Under one or both of these conditions, multivariate techniques might be necessary. Among these methods, principal components analysis (PCA) is the easiest to use (see Appendix 9.1). The objective of this method is to describe the size and shape of multivariate distributions and to reduce a large set of correlated variables to a much smaller set of uncorrelated, transformed ones. For a set of multivariate normally-distributed characters, the first principal component is aligned in the direction of the greatest combined variance among the characters, expressed by its eigenroot, the second in the direction of the next greatest variance, and so on (Appendix 9.1). Therefore, if source means or the among source correlation matrix are analyzed, the first principal component would describe the greatest differences among sources (and so on). This provides the rationale behind Campbell's (1984) use of principal components analysis of among-source correlation matrices.

As noted above, correlation matrices can be computed from source means or from source variance-covariance matrices. The former are the easiest to compute and are usually well-behaved in further analysis. However, principal components analysis is sensitive to large error variation or complex patterns, which can obscure geographic patterns (Wartenberg 1985). Thus estimates of means must be precise so that existing patterns can be detected. Correlation matrices from genetic variances and covariances also are not without problems. If correlations are above one, then negative eigenroots and R^2's can result (Appendix 9.1). A further shortcoming of PCA itself is that the principal components might not be aligned in the direction of greatest geographic divergence among seed sources (Gittins 1985, Wartenberg 1985).

Under the latter condition, extensions of the multivariate general linear model might be optimal. The simplest of these is canonical analysis (CA), which is the multivariate form of the one-way analysis of variance. Each trait is *weighted* by its ratio of between- to within-source variance; traits having high ratios of between- to within-source variances are given greater weight, as are pairs of traits having between-source covariances that are opposite in sign to the within-source covariances. For our purposes, the appropriate matrices would be the sums of squares and crossproducts among sources, **B**, and its appropriate error matrix (source x site, etc.), **W**. The eigenvectors of the matrix product, $\mathbf{BW}^{-1}$, are aligned in the direction of the greatest relative differences among sources. Their roots are the multivariate equivalent of the ratio of the between-to-within-group sums of squares. Programs like SAS Institute's (1985) PROC CANDISC will provide statistical tests of the multivariate differences, as well as output scores of seed sources for all canonical vectors. Such canonical scores are linear functions of the original traits for each canonical vector, as illustrated in Appendix 9.2. Scores can be plotted by pairs of canonical vectors, particularly the first two, so that source means can be examined for trends.

A more direct elaboration of canonical analysis is the multivariate analysis of variance or MANOVA. This analysis is conducted just as the familiar ANOVA would be, with the appropriate error terms, except that the traits are analyzed simultaneously. In SAS, for example, the analysis is initiated with the MANOVA statement under PROC's ANOVA or GLM (the former for balanced and the latter for unbalanced designs). Because the CANONICAL option under the MANOVA statement provides more useful information than the default, it should also be specified. An important limitation of most (or perhaps all) current statistical packages is that they do not compute canonical scores for any given hypothesis under MANOVA. Thus if scores for sources are desired, you must copy the canonical weights from the output and compute the scores yourself from source means as follows: Say the p by 1 vector of weights is $\mathbf{b}$ and the matrix of traits is $\mathbf{X}$ for p traits and n observations. Then the vector of n canonical scores is $\mathbf{b}'\mathbf{X}$ (Appendix 9.2).

The absence of pattern under visual examination or by linear regression and correlation is not a sufficient test of the hypothesis that patterns do not exist because the structure may be high order and multidimensional. Complex geographic structure requires more complex statistical methods. The favored method is multiple regression, whereby measures of source location, such as latitude, elevation, and slope, are regressed against each trait. Then series of curves of predicted values are plotted against geographical variables to determine optima and rates of change. Such regressions are an essential part of Campbell's (1986) and Rehfeldt's (1986) analyses of short-term tests, except that Campbell extracted principal components of the among-source genetic correlation matrix and then regressed them against models of geographic location.

One of the most important and difficult steps in regression analysis is the development and selection of the model. Everyone seems to have their own strategy; mine is to begin with simpler models and build to the more complex. As the complex (or higher order) terms are added to the model, the additional fit these terms bring to the model can be tested for significance. For example, if seed source latitude (LAT), longitude (LON), and elevation (ELEV) are the geographic terms of the model, the linear terms are LAT, LON, and ELEV, the quadratic terms are LAT^2, LON^2, $ELEV^2$, LAT*LON, LAT*ELEV, and LON*ELEV, and the cubic terms are LAT^3, LON^3, $ELEV^3$, LAT^2*ELEV, and so on. At each incremental test of higher-order terms, residuals should be plotted against the independent variables. These can indicate the need for either additional terms or transformations of the dependent or independent variables (Draper and Smith 1966). In addition, the adequacy of fit of the higher-order terms depends, in part, on the statistical significance of these terms, added to the model (Box and Draper 1987). In SAS, using PROC GLM, the importance of the higher degree model is tested on the mean squares combined from the Type I sums of squares attributed to individual higher-order terms (for eg., the combined sums of squares for the cubic terms).

When three or more basic geographic variables compose the model, higher order terms can result in extremely large, overfitted models. Although such models should be reduced in size by variable selection procedures, such as SAS' PROC STEPWISE, the resulting equations are primarily of value as reduced models, but not for purposes of describing the importance of geographic variables in the model: because of correlations among independent variables, slope coefficients in resulting equations are unreliable indicators of the importance of variables in the equations. Consequently, there will often be important differences in the composition of equations for separate dependent variables, but remarkable similarity in their response surfaces (eg. Rehfeldt 1988).

Models also can be extended to regressions of sources across sites, as in Kung and Clausen (1984). By adding linear and squared terms of say, site latitude, shifts in the sources that are maximum for each site can be estimated. If a relatively large number of well-distributed sites are tested, say six or more sites, optima for locations can be estimated by interpolation.

As noted above, sets of traits either can show the same fundamental pattern or correlations among traits can obscure patterns. Under conditions such as these, multivariate regression, commonly called canonical correlation analysis (CCA), is more efficient than generating regressions for each trait. With this method, development of the models is essentially the same as in multiple regression, except that there are two to many dependent variables rather than one. From a purely statistical viewpoint, no distinction is made between dependent and independent variables in CCA. But from a practical standpoint, one set is treated in the manner of dependent variables (as in metric traits) and the other as independent variables (as in location variables).

The theoretical justification for this approach is the same as for principal-components analysis, followed by regression of geography against the PC's, except that CCA is more straightforward. And from the canonical structure (see below), direct inferences are made about the original data. Also, random errors are usually positioned in the residual error of the regression, not in the model, as can happen with PCA.

To illustrate regression procedures in general and canonical correlation analysis in particular, I will summarize results from Aitken et al. (1987) and Aitken, Westfall, and Libby (ms in preparation). Their analysis extends Kung and Clausen's (1984) model, by incorporating multiple traits and three measures of source location. The data are from a range-wide provenance test of white fir (*Abies concolor*), concentrating on Pacific coast sources and planted at three locations in California: Camino, Blodgett, and Hoopa. Although the models developed are based on the three sites, I will focus on results of a Sierra Nevada site, Camino (latitude 38.75°N longitude 120.67°W), and the Coast Range site, Hoopa (latitude 41.80°N, longitude 123.55°W). Growth, morphological indices, and survival were measured over 20 years.

The analytical procedure is as follows: First specify dependent traits and the geographic model just as one would in any multiple regression. For example, a model describing two traits, Y_1 and Y_2, that includes plantation latitude (PLAT), source latitude (LAT), and source longitude (LON) might be:

$$a_1Y_1 + a_2Y_2 = b_1PLAT*LAT + b_2PLAT*LON + b_3PLAT*LAT^2 + b_4PLAT*LON^2 + b_5PLAT*LAT*LON.$$

This model focuses on the interaction between plantation latitude and source latitude and longitude. When a canonical correlation program like PROC CANCORR from SAS Institute, Inc. (1985) is used, the dependent and independent variables are specified separately as VAR and WITH variables, respectively. Also, the independent variables included in the model must be previously defined in the data.

The model above is oversimplified. In reality, the solution to the canonical equations would yield pairs of equations for the traits (Y's) and predictors (X's) for each canonical vector. The solution for the left side of the equation would result in a canonical score for the multivariate trait for each observation and canonical vector and, for the right side of the model, the score for the predictor (Eqs. 9.4 and 9.5, respectively, Appendix 9.2).

Box 9.1 shows the SAS job used to generate the results from the white fir example given in Box 9.2 and Table 9.2. This model specifies average effects of plantation latitude (PLAT) and longitude (PLON) and seed source latitude (LAT), longitude (LON), and elevation (ELEV), as well as the interaction between site location and source location.

The sequence of steps is as follows: First the trait data are entered by plantation and appended together. Some of the traits are transformed and plantation locations are also specified. Then source locations are entered, merged with the trait data, and the higher order terms of the trend surface model created. Then PROC CANCORR is run, limiting the output to six canonical vectors and calculating canonical scores to a temporary output file. To plot predicted scores and provide regression models for surface plots, PROC GLM is run on the canonical vectors and the predicted scores sent to another output file. Then contour plots of predicted scores are printed by plantation against source latitude and longitude (printout not shown).

The output of the CANCORR procedure provides unadjusted and unbiased canonical correlations, statistical tests of the canonical vectors, tables of canonical weights (the a and b slope coefficients, above), and correlations between the original variables and the canonical vectors. The SAS output for the correlations and significance tests is given in Box 9.2. Most of these data are either described in Appendix 9.2 or should be self-explanatory. However one term has not been defined elsewhere: the adjusted canonical correlation. It is an estimate of the unbiased correlation.

Box 9.1. SAS job to execute a CANCORR procedure

```
OPTIONS LS=80;
/*The following three data steps input the data from each planting site,
recode or create new variables, and drop unneeded variables. Each set is
sorted in preparation for merging.*/;
DATA CAMINO;
 LENGTH POP $ 6;
 INPUT PNAME $ POP $ MBN85 MH67 MS85 MH71 MH74
    MH83 MBHD83 MVI82 MCD MCF MSFL72 MS74;
 MH85=11*(MH83-MH74)/9;
 MBHD=MBHD83*MH85/MH83;
 MVI=MVI82;
 MBN=MBN85;
 DROP MSFL72 MH83 MBHD83 MVI82 MBN85;
*Plantation latitude (PLAT) and longitude (PLON);
PLAT=38.75;
PLON=120.67;
  CARDS;
;
PROC SORT; BY POP;
DATA HOOPA;
 LENGTH POP $ 6;
 INPUT PNAME $ POP $ MH67 MBN68 MH71 MH74 MH85
      MBHD MVI MCD MCF MSF MS74 MS85 MBN=MBN68;
PLAT=41.08;
PLON=123.55;
 DROP MSF MBN68;
 CARDS
;
PROC SORT; BY POP;
DATA BLODGE;
 LENGTH POP $ 6;
INPUT PNAME $ POP $ MBNK70 MH67 MH71 MH74 MH85
    MBHD MVI MCD MCF MSFL72 MS74 MS85;
    MBN=MBNK70;
```

(Cont'd next page)

(Box 9.1 cont'd)

```
 PLAT=38.90;
 PLON=120.60;
 DROP MSFL72 MBNK70;
 CARDS;
 ;
PROC SORT; BY POP;
*Input seed source locations;
DATA LOC;
 LENGTH POP $ 6;
 INPUT POP $ LAT LON ELEV;
 CARDS;
 ;
PROC SORT; BY POP;
*Append data from the three sites together;
DATA TOT;
 SET CAM HOOP BLODG;
/*Merge the planting site data with source locations, matched by seed
    source (POP). In addition, second order (eg LATSQ) and site by
    source location interaction terms (PLATLAT) are created*/;
DATA TOT;
 MERGE TOT LOC; BY POP;
 LATSQ=LAT*LAT;
 LATLON=LAT*LON;
 LATELEV=LAT*ELEV;
 LONELEV=LON*ELEV;
 PLATLAT=PLAT*LAT;
 PLALATSQ=PLAT*LATSQ;
 PLATLON=PLAT*LON;
 PLALATLO=PLAT*LATLON;
 PLALATEL=PLAT*LATELEV;
 PLALONEL=PLAT*LONELEV;
 PLATELEV=PLAT*ELEV;
 PLONLAT=PLON*LAT;
 PLOLATSQ=PLON*LATSQ;
```

(Cont'd next page)

(Box 9.1 cont'd)

```
 PLONLON=PLON*LON;
 PLOLATLO=PLON*LATLON;
 PLOLATEL=PLON*LATELEV;
 PLOLONEL=PLON*LONELEV;
 PLONELEV=PLON*ELEV;
*Invoke the canonical correlation analysis and output canonical
    scores into a work file (CAN);
PROC CANCORR NCAN=6 OUT=CAN /* Results for the first six
    canonical vectors are printed*/;
 VAR MH67--MBN;
 WITH PLAT--PLONELEV;
*Generate predicted scores for vectors 2 and 4 into work file PV;
PROC GLM;
 MODEL V2 V4=PLAT--PLONELEV;
 OUTPUT OUT=PV P=PV2 PV4;
PROC SORT; BY PNAME /*Sort by planting site name*/;
*Contour plots of predicted scores for vectors 2 and four by source
    latitude and longitude for each planting site;
PROC PLOT;
PLOT LON*LAT=PV2/CONTOUR=8 /* Eight contour intervals*/;
PLOT LON*LAT=PV4/CONTOUR=8;
  BY PNAME;
```

The first five vectors account for about 90% of the patterned variation (see the CUMULATIVE column). Excluding the first vector, which reflects site differences, the remaining vectors describe about 75% of the patterning. Keep in mind that the eigenvalues are the multivariate equivalent of the ratio of sums of squares due to the model versus the residual sums of squares. In the interest of avoiding unnecessary detail, correlations for the first five vectors are summarized in Table

Box 9.2. **Canonical Correlation Analysis**

113 Observations
11 'VAR' Variables
23 'WITH' Variables

Note: The correlation matrix for the 'with' variables is less than full rank. Some canonical correlations and coefficients will therefore be zero.

	Canonical Correlation	Adjusted Canonical Correlation	Approx Standard Error	Squared Canonical Correlation
1	0.961912	0.952063	0.007061	0.925275
2	0.806092	0.725920	0.033092	0.649784
3	0.772298	0.699366	0.038133	0.596443
4	0.738033	0.676210	0.043022	0.544693
5	0.699185	0.653660	0.048298	0.488860
6	0.618345	0.495459	0.058362	0.382350
7	0.593224	.	0.061238	0.351915
8	0.491806	0.407585	0.071636	0.241873
9	0.421417	0.354585	0.077710	0.177592
10	0.336636	0.253880	0.083783	0.113324
11	0.273317	0.251774	0.087432	0.074702

EIGENVALUES OF INV(E)*H = CANRSQ/(1-CANRSQ)

	Eigenvalue	Difference	Proportion	Cumulative
1	12.3824	10.5270	0.6262	0.6262
2	1.8554	0.3774	0.0938	0.7200
3	1.4780	0.2816	0.0747	0.7948
4	1.1963	0.2399	0.0605	0.8553
5	0.9564	0.3374	0.0484	0.9036
6	0.6190	0.0760	0.0313	0.9349
7	0.5430	0.2240	0.0275	0.9624
8	0.3190	0.1031	0.0161	0.9785
9	0.2159	0.0881	0.0109	0.9895
10	0.1278	0.0471	0.0065	0.9959
11	0.0807	.	0.0041	1.0000

(Cont'd next page)

(Box 9.2 cont'd)

Tests of HO: The Canonical Correlation in the current row and all that follow are zero

	Likelihood Ratio	Approx F	Num DF	Den DF	PR>F
1	0.00050327	4.7525	209	819.91	0.0001
2	0.00673496	3.2320	180	759.528	0.0001
3	0.01923084	2.9071	153	697.243	0.0001
4	0.04765340	2.5966	128	632.871	0.0001
5	0.10466216	2.2797	105	566.177	0.0001
6	0.20476228	1.9469	84	496.862	0.0001
7	0.33151841	1.7189	65	424.541	0.0009
8	0.51153562	1.3809	48	348.728	0.0553
9	0.67473589	1.1637	33	268.807	0.2548
10	0.82043944	0.9570	20	184	0.5164
11	0.92529768	0.8342	9	93	0.5865

MULTIVARIATE TEST STATISTICS AND F APPROXIMATIONS

S=11 M=3.5 N=40.5

Statistic	Value	F	Num DF	Den DF	PR>F
Wilks' Lambda	0.000503	4.752	209	819.91	0.0001
Pillai's Trace	4.546812	3.449	209	1023	0.0001
Hotelling-Lawley Trace	19.774	7.681	209	893	0.0
Roy's Greatest Root	12.38235	60.608	19	93	0.0001

Note: F statistic for Roy's Greatest Root is an upper bound

9.2. Of the canonical vectors in Table 9.2, the first reflects differences among sites, hence the high correlations with site location. The remaining four vectors reflect patterns in the traits of interactions

Table 9.2. Canonical loadings and rates of change for 5 canonical vectors for geographic patterns in white fir traits.

		1		2		3		4		5	
Variable	Mean	r[a]	b[b]	r	b	r	b	r	b	r	b
Traits											
BN	4.53	0.83	1.05	.	.	.	.	0.17	0.27	0.17	0.30
CF	0.85	-0.22	-0.11	.	.	0.19	0.11	-0.24	-0.15	0.16	0.11
S74	53.75%	0.47	10.90	-0.10	-2.91	0.20	5.77	0.40	12.26	0.15	4.88
S85	51.60%	0.54	12.30	-0.10	-2.68	0.21	6.06	0.37	10.93	0.18	5.68
H67	20.56cm	0.28	1.56	.	.	0.44	3.09	-0.19	-1.36	0.12	0.95
H71	77.89cm	0.41	9.51	0.16	4.38	0.48	14.00	.	.	0.35	11.13
H74	133.42cm	0.74	40.73	.	.	0.31	21.09	.	.	0.24	18.45
H85	344.70cm	0.37	43.18	-0.30	-41.43	0.31	44.45	0.35	51.96	0.17	27.07
DBH	8.14cm	.	.	0.13	2.85	0.37	8.49	-0.09	-2.24	0.10	2.47
VI	19.70	0.57	12.97	-0.36	-9.66	0.40	11.21	.	.	0.12	3.79
Geographic Model											
PLAT		-0.77		.		0.11		0.20		0.22	
PLON		-0.73		.		0.12		0.21		0.24	
LAT		-0.09		0.21		-0.44		.		.	
LON		0.09		-0.24		-0.22		0.36		-0.21	
ELEV		.		0.19		0.22		0.10		.	
LATSQ		-0.10		0.22		-0.45		.		.	

(Cont'd next page)

(Table 9.2 cont'd)

Variable	Mean	1		2		3		4		5	
		r[a]	b[b]	r	b	r	b	r	b	r	b
Geographic Model											
LATLON		.		0.13		-0.43		.		.	
LATELEV		.		0.32		.		0.14		.	
LONELEV		.		0.15		0.19		0.17		.	
PLATLAT		-0.30		0.21		-0.39		.		.	
PLALATSQ		-0.21		0.21		-0.43		.		.	
PLATLON		-0.48		-0.14		-0.09		0.41		.	
PLALATLO		-0.25		0.13		-0.39		0.11		.	
PLALATEL		-0.14		0.33		.		0.17		.	
PLALONEL		-0.08		0.16		0.21		0.20		.	
PLATELEV		-0.09		0.20		0.24		0.13		.	
PLONLAT		-0.18		0.22		-0.42		.		.	
PLOLATSQ		-0.14		0.22		-0.44		.		.	
PLONLON		-0.19		-0.21		-0.17		0.42		-0.11	
PLOLATLO		-0.13		0.13		-0.42		0.09		.	
PLOLATEL		.		0.32		.		0.16		.	
PLOLONEL		.		0.16		0.20		0.19		.	
PLONELEV		.		0.20		0.23		0.12		.	

[a] Canonical loading: the correlation between the variable and the scores of the opposite canonical variable. This is the multivariate equivalent of the correlation between an independent variable and the dependent variable.

[b]The unit change in the trait with a unit change in score for the canonical vector.

between site and source locations (patterns of rank changes of sources, or G x E interaction across the three sites). The second vector generally describes patterns of 22-year growth; the third, early growth; the fourth, survival; and the fifth, mid-term growth. Correlations between linear combinations of the traits and locations are 0.96, 0.81, 0.77, 0.74, and 0.70 for the first through the fifth vectors, respectively, and are all statistically significant (Box 9.2). These correlations are the multivariate equivalent of the multiple correlation between the dependent variable and the linear combination of the independent variables in regression. The square of this correlation is, of course, the coefficient of multiple determination or R^2. Because the most recently measured traits are of interest in defining seed zones, I will focus on the second (late growth and size, H85 and VI) and fourth (survival, S74 and S85) patterns.

Some important points regarding these results should be noted. First, because there have been age-related changes in patterns of growth, these patterns can then be defined in separate or independent, functions. Furthermore, the variation in each trait has been partitioned among all possible canonical vectors or functions. Thus the variation described by a given number of canonical vectors can be computed by summing the squared canonical loadings for these vectors (Appendix 9.2). For example, 47% ($0.37^2 + 0.30^2 + 0.31^2 + 0.35^2 + 0.17^2$ from the r's in Table 9.2) of the variation in 1985 height is described by the first five canonical functions. In addition, the trait values can be predicted from the functions. I have done this by regressing scores of each of the trait vectors against each of the traits. Because canonical scores are standardized to a mean of zero and standard deviation of one, the intercept of each regression is the trait mean and the slope coefficient (b, Table 9.2) is the change in the trait with a one standard deviation change in canonical score. Then sources that are one standard deviation below the mean for the second vector and one standard deviation above the mean for the third and fourth vectors would be about 140 cm greater than the mean for 1985 height (-1*(-41) + 44 + 52). Thus geographic patterns can be based on traits of interest, which in turn can serve to guide development of seed (procurement) zone boundaries.

As noted above, I generated regression functions for the two patterns using PROC GLM and using the same model for geographic traits as was specified in CANCORR (Box 9.1). This was done to generate predicted canonical scores. Then to examine patterns in these functions graphically, I iterated these functions over intervals of predicted seed source latitude and longitude for each of the two sites and for three source elevations, generating a matrix of predicted scores for each set of iterations. These predicted scores were generated and plotted using graphical functions from Mathematica™ (Wolfram 1988). The following are contour plots of the growth function (V2), predicted for source locations and three elevations at Camino (Fig. 9.2). A range map of *Abies concolor* within California is provided for reference (Fig. 9.3).

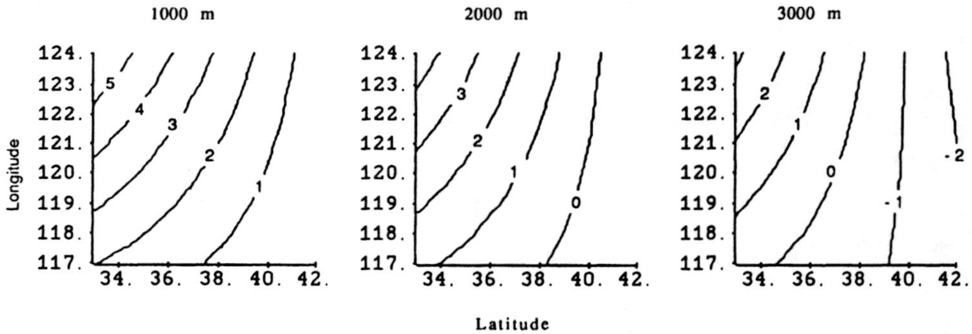

Figure 9.2. Contour plots for predicted scores of the growth vector at Camino for 1000, 2000, and 3000m elevation. Contour lines are in one standard deviation intervals.

Because 1985 height is negatively correlated with scores of this vector, I multiplied the scores by -1. Corresponding plots for the survival function (V4) are in Fig. 9.4.

The fastest-growing sources are at lower elevations in the south, whereas the best surviving sources are those at higher elevations in the north. Because source elevations range from about 1800 m in the northern Sierra to about 2000 m in the southern Sierra and then about 1600 m in the Transverse Ranges, growth maxima extend from local sources to those of the Transverse Ranges. Maxima in survival center around local sources at the higher elevations.

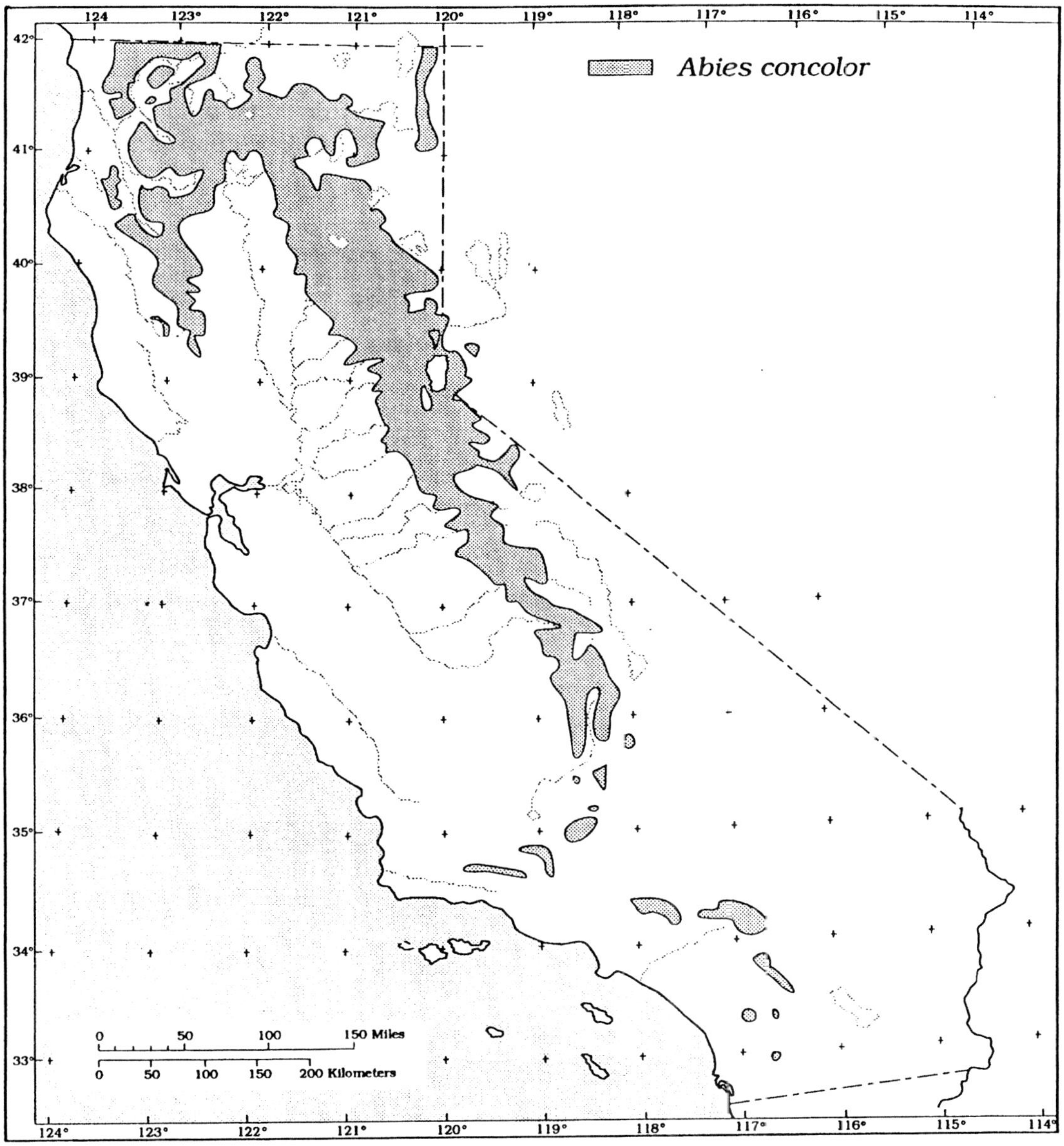

Figure 9.3. Range map of white fir *(Abies concolor)* in California.

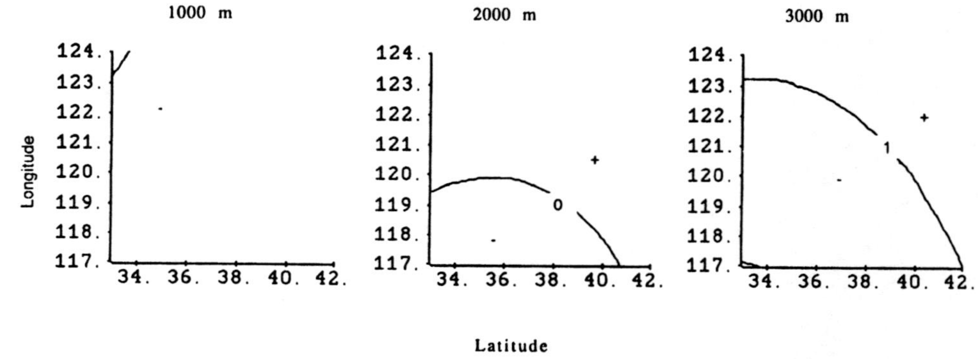

Figure 9.4. Contour plots for predicted scores of the survival vector at Camino for 1000, 2000, and 3000m elevation. The 0 contour line for 1000 m is at the upper left. Regions above and below a contour line are indicated by + and -, respectively.

Plots of predicted scores for the growth function at Hoopa are shown in Fig. 9.5.

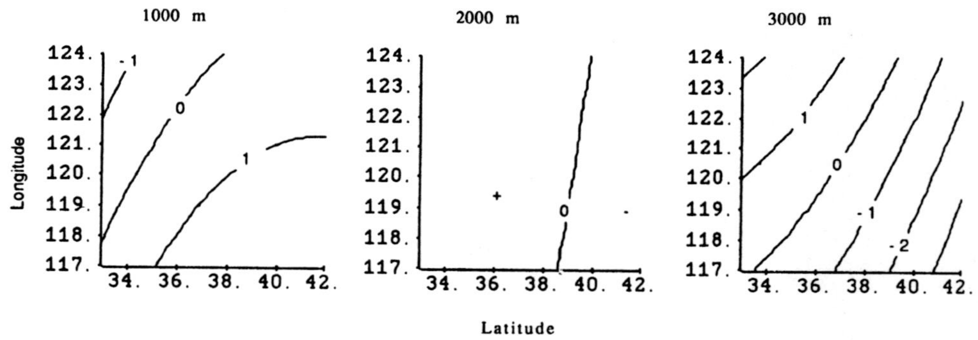

Figure 9.5. Contour plots for predicted scores of the growth vector at Hoopa for 1000, 2000, and 3000m elevation. Regions above and below a contour line are indicated by + and -, respectively.

Plots for the survival function are shown in Fig. 9.6.

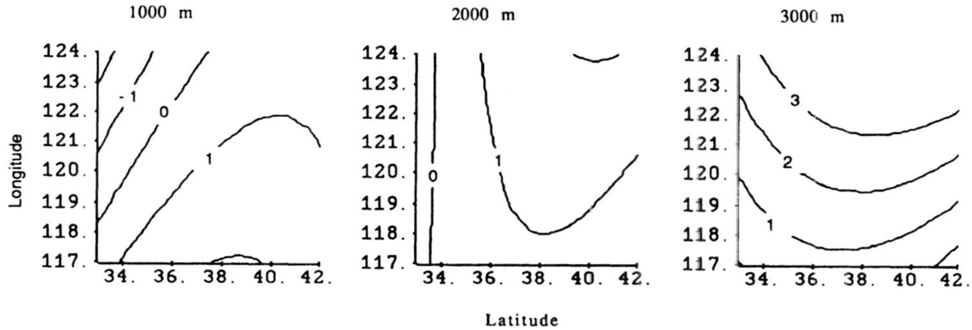

Figure 9.6. Contour plots for predicted scores of the survival vector at Hoopa for 1000, 2000, and 3000m elevation.

For growth, the best sources are at lower elevations in the northern and mid to higher elevations in the southern Sierras. In contrast, survival is highest in the mid to southern Sierras at the lower elevations, and on the North Coast at upper elevations. For the sampled sources, the fastest-growing sources would be centered at about 38°N, while the best surviving are to the north and west of this latitude.

Series of surface or contour plots such as those in Figs 9.2-9.5 above are admittedly difficult to mentally integrate and visualize. Instead, scores can be predicted for the sampled sources and graphed in 3D surface or contour plots (as done in Box 9.1). Or more simply, scores above a given level, say one or two standard deviations, can be selected and listed after removing data that are outside the species' range. Indicators of transfer risks in various directions can be estimated from canonical analysis of the equations (Box and Draper 1987).

The procedures above weight each variable in the analysis equally and do not consider its relative value. When the value of each trait is important, modifications of regression equations are required. One alternative is to develop separate regression equations for each trait, apply relative weights to the equations, and combine the equations to determine optimum source locations from the weighted function (Namkoong 1979:187-194). In a second alternative, scores of canonical vectors can be weighted according to the traits composing the vectors and the scores summed together and plotted to establish maxima. For example, if survival is ten times more important than 20-year height, the combined score might be: 10V4 - V2 (1985 height is negatively correlated with V2). The weighting will thus shift the combined maximum toward the local sources.

Regression analyses are not without their drawbacks. First, the data may not be well described by models using the geographic variables available. Similarly, discontinuous patterns tend to be smoothed out by regression, unless higher order equations are used. Finally, the data can be easily overfitted, especially with canonical correlation analysis (CCA), whereby the number of dependent plus the independent variables can constitute a substantial portion of the total degrees of freedom. Although variable selection procedures are not available for CCA, models can be developed by sequentially adding more complex terms, and checking significance levels of canonical vectors after each addition. Then the models for individual canonical vectors can be examined diagnostically by univariate procedures on canonical scores. Because of the addition of dependent variables to the model, CCA also requires larger sample sizes to avoid overfitting. Rules for ratios of sample sizes to numbers of variables have not been developed; I tend to prefer five times as many samples as the total number of variables, dependent and independent. However, a recent study of the behavior of multivariate techniques suggest that a 5:1 ratio is more than needed, except where sample sizes are smaller than 100 (Williams and Titus 1988). Additional analyses have shown that, except where patterns are highly complex, fewer traits than are often reported in the literature are needed to effectively describe patterns (Thorpe 1987).

A non-linear alternative to the above-described linear models has been developed by Raymond and Lindgren (1986). The method first requires that measures of the site and the source be reduced to respective indices. Then the Cauchy equation is solved by nonlinear regression:

$$Y = \frac{a}{1 + \left[\frac{X - b}{c}\right]^2},$$

where:

- Y = site or provenance index
- X = relative performance of the site or provenance (max = 100)
- a = maximal performance
- b = the value of the optimum
- c = the width of the response curve at Y = 0.5 a.

The equation finds optima for either provenances to sites or sites to provenances. Where optimum sites are desired, then X = sites and Y = provenances, and conversely for optimum provenances. The theoretical maximum for 'a' is 100. The width, c, is an index for transfer risk.

Admittedly, parameters of the Cauchy equation provide much more information than do parameters of regression equations (excepting canonical analysis, described in Box and Draper 1987). But I have doubts about the broad applicability of the equation to the establishment of procurement zones. Perhaps the most important objection I have with the method is its requirement for a single index of the environment both in terms of site and source locations. This is a particularly serious limitation in areas of species diversity and site heterogeneity because, with our large number of species and sites, we do not have the experience needed to reduce source and locations to develop indices for each species in our breeding programs. Furthermore, there is no opportunity to determine the validity of the index values and the model assumes that the values have the same form over all traits selected. Finally, the function treats sites and provenances as fixed effects and does not provide information about the nature of genotype by environment interaction. Presumably, though, provenance (source) indices can be mapped. Thus sites sharing similar provenance optima (b's) can be clustered according to an acceptable width (c). These sites would compose a deployment zone and corresponding sources the procurement zone.

To summarize, seed zones can be set on the basis of single traits by tabulating or plotting seed source (provenance) means against source locations. Particularly if source populations were collected along a single geographic transect, trends should be readily apparent. It is also prudent at this stage and before proceeding further, to have statistically tested differences among source means. Better, the top-ranking say, 10% of the sources should be tested against the remainder. In SAS, this test is accessed by the CONTRAST option under the GLM procedure (see Ch. 6 for an example). Nonsignificant differences among source means result from one or both of two causes: small differences among the sources or large errors. The former condition, low variances among sources, suggests that zonation is unnecessary and so is further analysis. The latter condition can arise from either large genetic variances within sources or large experimental errors. Large genetic variance within sources will reduce transfer risks and the number of breeding zones estimated under the methods discussed in Section 9.2.3. Under the large experimental errors, analysis should also proceed, but with the realization that these substantial errors may be imbedded in the results.

If sources were sampled along many location criteria, strong patterns can still be detected by contour plots of source means against two measures of geographic location, as specified in the PROC PLOT statements in Box 9.1. Even when multidimensional patterns are readily seen, the data should be fitted by multiple regression. By examining contour plots generated from a regression model, you will be better able to visualize existing patterns, which, in turn, will aid in the development of transfer guidelines. In developing such regression models, the adequacy of first-, second-, and third-order models should be tested. The third-order model is a reasonable limit because experience to date suggests that it is unlikely that higher-order models will be necessary; third-order models comprising three or more environmental variables are complex enough.

Analyses of single traits are sufficient when only one trait is of interest or patterns of many traits are independent of one another. When neither condition holds (the latter is highly unlikely), multivariate extensions of the methods above should be employed. Plots of source means are extended to multivariate dimensions by the principal components analysis of the among-source correlation matrix and plotting scores of the principal components. Similarly, scores from the canonical analysis (CA) of family means in sources are plotted against source location. Canonical analysis, the multivariate equivalent of a one-way ANOVA, is usually more effective in differentiating among populations than PCA.

PCA and CA are useful when geographic patterns are simple. But when linear or stepped clines or ecotypic clusters are not obvious, or even when curvilinear trends are evident, multiple regression or its multivariate counterpart, canonical correlation analysis (CCA), should be utilized to test the existence of pattern. As a matter of course in the analysis, the sufficiency of lower-order models should be tested along with the adequacy of alternative models (Box and Draper 1987). For the latter, plots of residuals and tests of lack-of-fit are essential. Once a model is selected, then plots of the functions are useful for visualizing geographic patterns, choosing optimum locations, and assessing transfer risks. If you feel comfortable with the statistical methods and feel adventuresome as well, consider canonical analysis to the regression equations (Box and Draper 1987). Canonical transformation of equations, which is alike in principle but differs in form from CA, above, provides useful descriptions of the form of patterns and numerical estimates of transfer risks.

9.2.3 Choosing optimum seed sources: Contending with transfer risks in a finite number of seed zones

Theoretically we might prefer to be able to match each site with a set of maximally performing seed sources. We would sit at our computer terminals, supplied with equations such as those presented above, specify the site, and be given a list of sources suited to our objectives. This approach treats source selection on a sliding scale, much as that described by Rehfeldt (1986, 1990), whereby the optimum sources shift with a corresponding shift in planting site location. However, breeders are currently limited to assembling sources into defined units (procurement zones), and the propagules of these deployed from a fixed set of seed orchards or clonal production facilities. Thus we must define a region small enough to minimize the transfer risk among locations within each region, but large enough to be managerially practical. In what follows, I will discuss two approaches described in the literature, one that estimates transfer risk along environmental gradients, developed by Campbell (1986), and the other, which maximizes expected value in an environmental gradient, proposed by Roberds and Namkoong (1986). Both are based on the same concept, the ratio of genetic variance to variances among environments, and both making use of available data from seed source-tests.

The first of these is Campbell's (1986) concept of transfer risk. Geographic patterns are described for each trait or set of traits, in this case, principal components (PC's). For each PC, the additive genetic variance within source location is estimated from the pooled variances among pairs of females over some proportion of the locations. Using this information, the number of seed zones is set by a minimum probability of mismatch within zone, defined as follows. The additive variance within a source location describes a normally distributed set of genotypes matching a set of microsites at the location, illustrated by the curve on the left in Fig 9.7. Thus the set of microsites is itself normally distributed and also the correlation between genotypes and equivalent microsite environments is equal to one. Now let us say that this set of genotypes is moved some distance, x, away from its native, or optimal, environment to another (Fig. 9.7). The proportion of overlap, V, between the genotypes of one environment (the left distribution in Fig. 9.7) and the genotypes of another (the right curve in Fig. 9.7) is defined by

$$V = 1 - 2\alpha \qquad \textbf{Eq. 9.1}$$

where:

V = the proportion of overlap between genotypes and sites and

α = $P[0 < Z < (x/2)/\sigma_A]$ or the area under the standard normal curve from 0 to $(x/2)/\sigma_A$,

and:

x = the difference between the mean of the source and that of a source at another site (Fig 9.7)

σ_A = the standard deviation of the additive genetic variation within sources.

$(x/2)/\sigma_A$ = the standardized half-distance between genotypes and microsites.

Z = the standardized canonical score.

The areas defined by $0.5 - \alpha$ and by V are shown by the shaded portions in Fig. 9.8, a and b, respectively. The area to the left of x/2 in V (Fig 9.8b) means that although there are many genotypes available at a microsite, there are inadequate numbers of these microsites available. The risk in transfer (R) is then

$$R = 1 - V = 2\alpha.,$$

or the shaded area in Fig. 9.8c.

To illustrate, let us say that genotypes at 1000 m elevation are represented in curves on the left in Figs. 9.7 and 9.8. These genotypes are to be planted at 1500 m; the distribution of genotypes and their

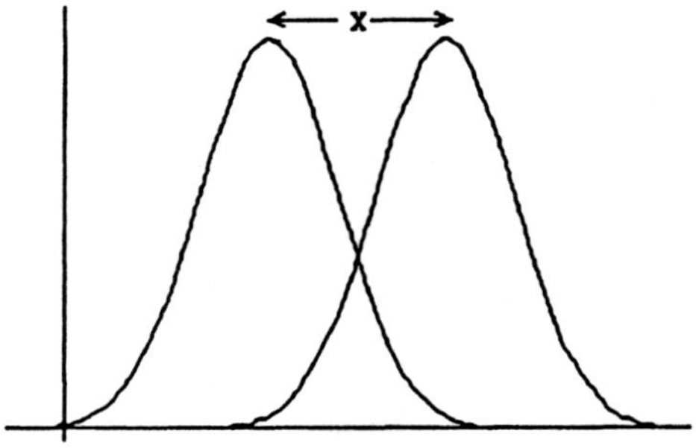

Figure 9.7. The distribution of genotypes at one location (the curve on the left) and genotypes at another location along an environmental gradient (the curve on the right). The difference between genotypes at the two locations is given by x and is 2.6 standard deviations.

matching microsites at this elevation is shown in the curves on the right of the same figures. Thus x = 500m and, with this difference being 2.6 standard deviations, σ_A = 192.3m (keep in mind that the trait is scaled in terms of the environment in this example). The area under α or $P[0 < z < (500/2)/192.3]$ is 0.403 (Table 11, Rohlf and Sokal 1981). Thus $V = 1 - 2\alpha = 1 - 2*0.403 = 0.19$ or 19% of the genotypes at 1000m match microsites at 1500m. And the proportion of genotypes at 1000m not matching microsites at 1500m is $R = 2*0.403 = 0.81$.

When multiple traits are the basis for zone boundaries, a then depends on the multivariate normal probability distribution of the traits and is given by simultaneous shifts in all characters (Fig 9.9). In PC's and canonical vectors, which are uncorrelated, combined risk of transfer for two vectors is as above except that:

$$\alpha = \Pr[0 < \Delta < 1/2\delta]$$

where,

$$\delta^2 = (\bar{x}_1 - \bar{x}_2)'S^{-1}(\bar{x}_1 - \bar{x}_2), \qquad \textbf{Eq. 9.2}$$

and,

$\bar{x}_1$ and $\bar{x}_2$ = the vectors of means for genotypes at locations one and two, respectively, and,

S = the additive genetic variance-covariance matrix.

Note that $1/2\delta$ is the multivariate equivalent of $(x/2)/\sigma_A$.

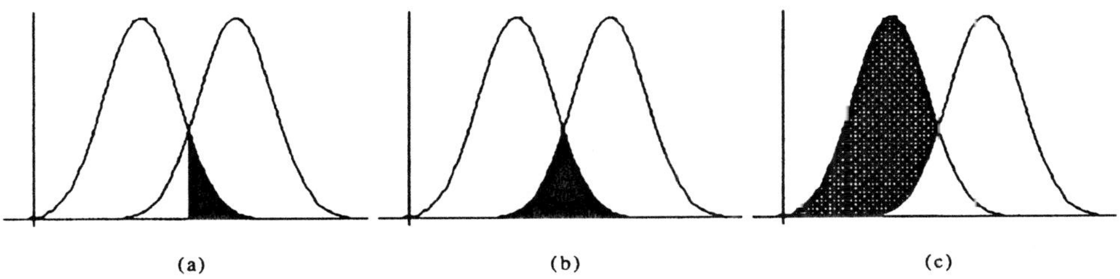

Figure 9.8. Areas of overlap and mismatch between genotypes of one environment and genotypes of another along an environmental gradient. The shaded areas define (a) 0.5 - α, (b) the overlap between genotypes and microsites, V, and (c) the area of mismatch or transfer risk, R.

To carry the illustration in Fig. 9.9 forward, the combined risk of transfer for two vectors, assuming a 2.6 standard deviation transfer in each, $1/2\delta = 1.84$, meaning that $R = 0.93$. When the x_p's are in terms

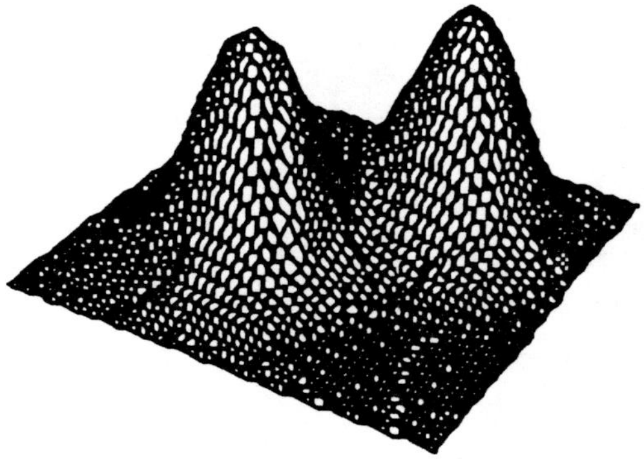

Figure 9.9. Bivariate normal probability distributions for genotypes at one site (lower left) and those at another (upper right). Means for the characters represented on both the x and y axes are 2.6 standard deviations apart.

of standard deviations, the above can be simplified to

$$\delta^2 = x^2_1 + x^2_2 + x^2_3 + \ldots + x^2_p,$$

for p vectors. Thus for 2.6 standard deviation differences in three vectors, $1/2\delta = 2.25$ and $R = 0.98$.

If you are familiar with multiple trait selection, you will see from (2) that δ depends on the correlations among traits. Thus when traits are positively correlated, δ decreases as ρ increases to a minimum of the mean standard deviation difference for all traits. But when traits tend to be negatively correlated, δ increases in value.

Equation 9.2 is taken from misclassification probabilities in discriminant analysis (Morrison 1976:235). This differs from Campbell's (1986) method for computing combined mismatch risks in that the risk is

$$CR = R_1 + R_2 - R_1R_2, \qquad \textbf{Eq. 9.3}$$

where R_1 and R_2 are the risks for the first and second vectors, respectively. This equation (9.3) can be shown to overestimate combined risks for uncorrelated traits by specifying an area under the bivariate normal curve that is greater than that expected for two overlapping curves. Moreover, Eq. 9.3 is invariant with correlations among characters.

Probabilities computed in the manner of Eq.s 9.1 or 9.3 assume that all genotypes within a population are maximally adapted to some environment within the site. Another crucial assumption is that of no site by seed source interaction in response functions. If such an interaction exists, then the method is appropriate only for projecting transfer risks over limited distances from the planting site.

The number of zones can be set by a minimum proportion of mismatch in the genotypes within each zone. From Eq. 9.2, steeper response functions (that is, resulting in a larger **x**) will increase the number of zones needed. Conversely, if additive variance within locations is large relative to changes in a surface (smaller δ), fewer zones will be necessary. The method can also be applied to artificially constructed seed lots, but these must be tested along with the native seed lots in order to standardize variances.

A second approach for determining optimum sources along an environmental gradient has been proposed by Roberds and Namkoong (1986, 1989). As in the above, the genetic variance of a trait is assumed to be normally distributed along an environmental gradient. This variance is defined in Roberds and Namkoong (1989) as the measure of uniformity or homeostasis of response over an environmental range, η_i, for the i'th population. Also, as in the above, the environment is assumed to be a normally distributed variable; that is, the set of environments used has a normal distribution. Also, the "response function" of the trait is scaled as a function of the environment. For a finite number of seed zones, mode source locations are found that maximize an expected value of the response function, given a ratio of genetic response to environmental variances. For two and three seed zones, lower-mode locations of seed sources of various ratios of η_i to the variance among environments, σ_e^2, (the ratio is specified as the homeostatic index, hs, in Roberds and Namkoong 1989) are given below in Table 9.3 (from Table 2, Roberds and Namkoong 1986).

Source locations are in terms of standard deviations of environmental units from the mean environment and zone boundaries are midway between mode locations. For two zones the latter would be the mean environment along the transect. And for three zones, the mode source location for the center or second zone would be at the mean environment.

Table 9.3. Mode locations (in standard deviations of environmental units) for zones under ratios of η/σ^2_e (hs).

hs	2 Zones	3 Zones
0.01	-0.24	-0.43
0.10	-0.49	-0.82
1.00	-0.72	-1.13
10.00	-0.79	-1.21
100.00	-0.80	-1.22

When the ratio of genetic (within population) to environmental variance is very low (0.01), optimum source locations are close to one another. Under such a ratio, two zones would increase the expected value of the response over one zone by 93% and three zones would increase the value over two by 43% (Roberds and Namkoong 1986). If, on the other hand, the ratio is high, locations are farther apart, the distance becoming asymptotic at hs of about 10. At a ratio of 10 or greater, two zones would be sufficient.

This method was applied to the elevational transect data from Conkle (1973). The variance among elevations of the ith source for the height response function was computed by:

$$s_i^2 = \frac{\sum_{j=1}^{e} \bar{x}_{ij} y_j^2}{\sum_{j=1}^{e} \bar{x}_{ij}} - \left[\frac{\sum_{j=1}^{e} \bar{x}_{ij} y_j}{\sum_{j=1}^{e} \bar{x}_{ij}} \right],$$

where,

$\bar{x}_{ij}$ = the mean height of the i'th source at the j'th site,
y_j = the elevation at the j'th site, and
e = the number of elevation zones.

The estimate for η was equal to the square root of the average of all the s^2_i's or 580. For a mean site of 915 m elevation, and an elevational variance one half of that for s_g^2 (ie, σ_e = 435 m), the lower mode, 1.062 standard deviations below the mean site, was at 480 m and could be planted up to 695 m. The middle zone had a mode of 915 m and could be planted between 695 and 1135 m, while at the upper zone, seed collected from 1350 m could be planted above 1135 m.

Application of this method is limited by the form of the distribution of the response function: it must be approximately normal. However, in the elevational transect data, variances in the response function were surprisingly equal among sources, considering the complexity of the response (Fig 9.1) and the limited number of sites. Under the assumption of normal (Gaussian) functions, mode locations are symmetrically distributed around the environmental mean, μ_e, although local sources are not required to be optimal (Roberds and Namkoong 1989). Another limitation in the usefulness of the method is that it is restricted to a single environmental gradient, although it has been recently extended to two (J. Roberds, personal communication). However, complex, multidimensional gradients could be reduced to a few independent ones by CCA and the above method applied to the transformed data for each of the vectors. The major difficulty here would be to effectively express the results in units of the original variables.

Conceivably, the method can be applied to functions of trait response to environmental gradients other than Gaussian, such as the regression functions discussed in Sect. 9.2.2. However, in the white fir data, 'optima' for source latitude, longitude, and elevation are outside the geographic range of the species. Thus applicability of the method will depend on the data and the response functions ascribed to the data.

In summary, Campbell's risk of transfer accounts for genetic variances within as well as among sources, but is less elegant than the Roberds/Namkoong (RN) method and only considers performance at a single site. The RN method does not consider within source variation and assumes that the response function has a normal distribution. The Cauchy function (Sect. 9.2.2) also can be applied where the goal is to minimize risk by choosing the source with the maximum yield at an arbitrarily determined site and then finding the range of sites where yield falls within some minimum percentage of the maximum (Raymond and Namkoong 1990).

9.3 Transfers Within Breeding Zones

When seed zones are established, thus forming breeding zones, we cannot assume that genotype by environment interactions are nonexistent. In some cases, they are, but we might generally expect interactions as a consequence of microgeographic site differences,

differences in cultural treatments, and effects resulting from other traits, such as insect and disease resistance. To make selections while taking interactions into account, it is often necessary to jointly describe the response of genotypes such as half- or full-sib families to environments. Moreover, the models used must effectively describe these responses according to the objectives of the breeder. Ultimately we might also desire models that describe the nature of the response and of any interactions that might exist. The simplest of these is the linear response function, presented in section 9.3.2. Introduced at this point is the concept of stability of genotypic performances over environments. Unfortunately, the definition of the stable genotype depends on the preferences of the individual breeder; there is no statistical or biological rationale for choosing one definition over another. Nor is there a statistical basis for choosing one method over the others.

Although linear models enjoy some popularity in forestry and are simple extensions of ANOVA, they may not fit the data; response functions may be intrinsically nonlinear. Such nonlinear functions will be discussed in section 9.3.3.

9.3.2 Linear functions - joint regression analysis

Concepts of adaptability and stability have been developed by Findlay and Wilkinson (1963) and Eberhart and Russell (1966). Findlay and Wilkinson (1963) partitioned G x E by regressing family against mean performance over environments by the model

$$Y_{ij} = b_{0i} + b_{1i}X_j$$

where,

Y_{ij} = the response of the ith genotype (eg, family) the jth environment (eg, site or plantation)
X_j = the mean response at the jth environment.

Thus the regression slope of individual families is compared to the average slope of all families over the environmental gradient. Their definition of the stable genotype is one that has a slope of one, that is, parallel to the average environment, and an adaptable genotype is one that has a high value for an intercept. Genotypes with a slope close to zero are regarded as highly stable. Eberhart and Russell (1966) extended this concept further by comparing the residual variance of regression for each parent. But in this case, high variance indicates low stability and a slope of one indicates high stability. As an example, these parameters were computed for five ponderosa pine families grown on four sites in the Central Sierra Nevada in California (J. Kitzmiller[3], unpublished results; Table 9.4). Families were selected for illustration on the basis of their contrasting values. Parameters for the linear equations are presented on the left side of the table: LSIG is the residual variance from the regression or the Eberhart and Russell (1966) stability parameter, and B_0 and B_1 are the Findlay and Wilkinson (1963) intercepts and slope coefficients, respectively.

Plots of the fitted equations, the function for the site averages, and the site means, are shown in Fig. 9.10.

---

[3]*J. Kitzmiller, Regional Geneticist, Region 5, USDA Forest Service, Chico, California*

Table 9.4. Parameters and residual variances from linear regressions[1] of mean environments against family means.

FAMILY	LSIG	B_0	B_1
3001	20.25	9.31	1.16
3049	0.45	-19.29	1.12
3350	11.12	-37.08	1.13
3357	10.36	48.83	0.76
3415	18.09	-20.18	0.95

[1]LSIG is the residual variance from the regression or the Eberhart and Russell (1966) stability parameter, and B_0 and B_1 are the Findlay and Wilkinson (1963) intercepts and slope coefficients, respectively.

Family 3001 has the highest slope coefficient and one of the highest residuals of the 45 families in the series of tests. However, it is among the top four families (it is the topmost curve) and is one of the most consistent performers of the set. In contrast, family 3415 is one of the poorest performers in all tests. In common with agricultural experience, there is a general tendency for the best families to have high slopes ($r = 0.62$) and a weak tendency for higher intercepts ($r = 0.35$). But, in exception to agricultural data (Lin et al. 1986), there was no association between family height and residuals in these data.

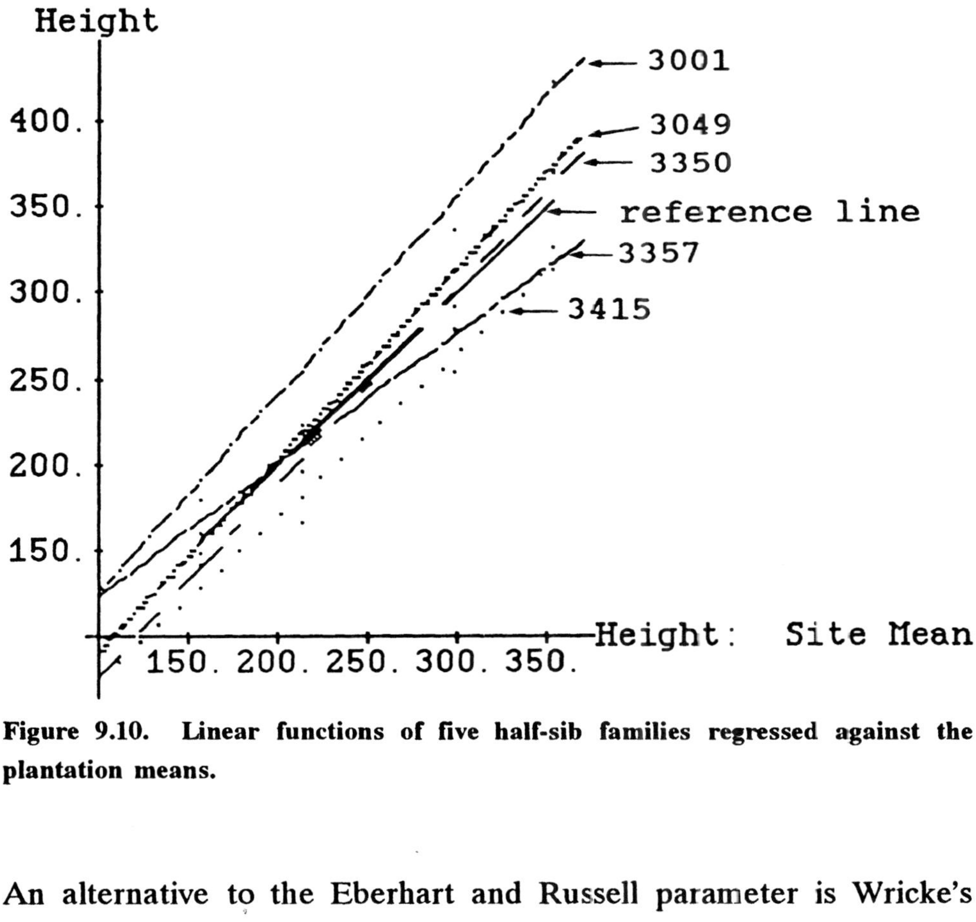

Figure 9.10. Linear functions of five half-sib families regressed against the plantation means.

An alternative to the Eberhart and Russell parameter is Wricke's (1962) ecovalence. This is:

$$\sum_{j=1}^{s} (\bar{X}_{ij} - \bar{X}_{i\cdot} - \bar{X}_{\cdot j} + \bar{X}_{\cdot\cdot})^2$$

where,

$\bar{X}_{ij}$ = the mean of the ith family on the jth site,
$\bar{X}_{i.}$ = the mean of the ith family,
$\bar{X}_{.j}$ = the mean of the jth site, and
$\bar{X}_{..}$ = the grand mean.

Ecovalence represents the contribution of each family to the interaction variance and thus is really an ANOVA parameter. It differs from Eberhart and Russell's residual variance in that ecovalence is not dependent on the fit of any model to the data (Lin et al. 1986). For the five families above, the ecovalences are given in Table 9.5.

There is a general trend in these data for families with high ecovalences to also be high in residual variance (r = 0.80). Thus the ecovalence tends to be more associated with heterogeneity of variances among sites than with heterogeneity of slopes. The SAS job used to generate the data in Tables 9.4 and 9.5 is shown in Box 9.3 below. Also included in the job is an analysis using a quadratic function to be discussed later. Shukla (1972) makes further extensions of ecovalence and linear slope parameters, but Kang et al. (1987) have shown that the stability-variance parameter, σ_i^2, is equivalent to ecovalence. A SAS program that computes these parameters is available from Kang (1989).

Table 9.5. Wricke's ecovalences for families in Table 9.4.

Family	Ecovalence
3001	1399.07
3049	326.79
3350	655.57
3357	1478.82
3415	704.29

These linear models have several shortcomings. First the regression functions are not independent: family means compose the site mean. Thus the parameters cannot be generalized unless the families tested can be considered as representing a random sample of all families that will be planted on the sites (Lin et al. 1986). Moreover, the repeatability of stability parameters is poor (Léon and Becker 1988). Also these parameters do not provide information on nature of genotype x environment interactions. Finally there is the assumption that G x E effects are additive; however, response functions may be nonlinear. As a result of the above, the utility of regression methods in breeding programs has been questioned.

For example, on examination of the data, responses for some of the families on the ponderosa pine data appeared to be nonlinear. To explore this possibility, I regressed a second order equation of plantation height against family height. The model I used is:

$$Y_{ij} = b_{0i} + b_{1i}X_j + b_{2i}X_j^2.$$

Box 9.3. SAS job computing joint regression analysis and stability parameters.

```
        OPTIONS LS=80;
        DATA GPM;
         SET DATASET.GPM;
         FAM=COMPRESS(FAM,'M');
         IF FAM='L' OR FAM='C705' OR FAM='C809' OR FAM='G531'
           THEN DELETE;
        PROC SORT; BY PLTN FAM;
*Compute means by plantation site and family;
        PROC MEANS NOPRINT;
         VAR HT;
         BY PLTN FAM;
         OUTPUT OUT=MNPF MEAN=HT;
        PROC SORT; BY PLTN;
*Compute plantation site means;
        PROC MEANS DATA=GPM NOPRINT;
         VAR HT;
         BY PLTN;
         OUTPUT OUT=MNP MEAN=PHT;
        DATA MNPF;
         MERGE MNPF MNP; BY PLTN;
        PROC SORT DATA=MNPF; BY FAM;
*Compute family means;
        PROC MEANS DATA=GPM NOPRINT;
         VAR HT;
         BY FAM;
         OUTPUT OUT=MNF MEAN=FHT;
         DATA MNPF;
         MERGE MNPF MNF; BY FAM;
*Compute test mean;
        PROC MEANS DATA=GPM NOPRINT;
         VAR HT;
         OUTPUT OUT=MNT MEAN=THT;
*Ecovalences;
        DATA MNPF;
```

(Cont'd next page)

(Box 9.3 cont'd)

```
    RETAIN THT;
    IF _N_=1 THEN SET MNT;
    SET MNPF;
    VH=(HT-FHT-PHT+THT)**2;
   PROC SORT; BY FAM;
   PROC MEANS NOPRINT;
    VAR VH;
    BY FAM;
    OUTPUT OUT=EV SUM=VH;
    TITLE "ECOVALENCES";
   PROC PRINT;
    VAR VH;
    ID FAM;
   DATA MNPF;
    SET MNPF;
    PHSQ=PHT*PHT;
*Finlay-Wilkinson/Eberhardt-Russell parameters;
   PROC REG DATA=MNPF OUTEST=EST1 NOPRINT;
    MODEL HT=PHT;
    BY FAM;
   DATA EST1;
     SET EST1;
     LINT=INTERCEP;
     LSIG=_SIGMA_;
*Quadratic parameters;
   PROC REG DATA=MNPF OUTEST=EST2 NOPRINT;
    MODEL HT=PHT PHSQ;
    BY FAM;
   DATA EST2;
     SET EST2;
     QINT=INTERCEP;
     QSIG=_SIGMA_;
     LHT=PHT;
     DROP INTERCEP _SIGMA_ PHT;
   DATA COMB;
    MERGE EV EST1 EST2; BY FAM;
   PROC PRINT;
    ID FAM;
    VAR LSIG LINT PHT QSIG QINT LHT PHSQ;
```

Parameters for these equations are also given in Table 9.6: QSIG is the residual variance of each equation and B_0, B_1, and B_2 are the intercepts, and slope coefficients for the linear and quadratic terms, respectively. These are plotted in Fig. 9.11.

Families 3350, 3357, and 3415 thus appear to respond quadratically to the four sites. Family 3357, with a positive linear parameter and negative quadratic parameter, is one that appears to perform better on the poor to moderate sites, but responds less well to the better sites. Conversely, family 3415, with parameters opposite in sign, responds efficiently to the more productive sites. The quadratic equation did not reduce the mean residual variance, in spite of the overfitting (there are three estimated parameters and four sites). Family height was not correlated with any of the slope parameters and ecovalence had a much lower correlation with the residuals ($r = 0.66$). Families having high residuals from the linear equation also tended to have large residual deviations from the quadratic ($r = 0.77$). Another noteworthy result was the high correlations among B_0, B_1 and B_2: as the linear term increased in value, the intercept and quadratic parameters decreased in value. These data were taken from a study of resistance to gouty pitch midge attack (Bedard and Westfall, unpublished data) and the second-ranked site was heavily infested with the midge and relative family differences were the largest among all the sites. Susceptibility to midge attack is highly heritable and the expectation would be that family means for midge infestation would be negatively correlated with B_1. Although this was the case, the correlation was nonsignificant. Thus other effects are responsible for this set of relationships.

Table 9.6. Parameters and residuals of quadratic regressions[1] of mean environment on families in Table 9.4.

FAMILY	QSIG	B_0	B_1	B_2
3001	28.11	-31.51	1.51	-0.00069
3049	0.16	-24.05	1.16	-0.00008
3350	1.47	79.41	0.13	0.00197
3357	5.67	-51.69	1.63	-0.00170
3415	6.07	164.75	-0.64	0.00313

[1]QSIG is the residual variance of each equation and B_0, B_1, and B_2 are the intercepts, and slope coefficients for the linear and quadratic terms, respectively.

A key question at this point is whether significant changes in rank occur (Baker 1988). For linear models such as those above, curves will intersect within an environmental variable greater than zero if $b_{ki} \neq b_{ki'}$ and $b_{k'i} \neq b_{k'i'}$, for two families i and i′ and two model parameters k and k′. Furthermore the two parameters must differ so that $b_{ki} < b_{ki'}$ and $b_{k'i} > b_{k'i'}$ or $b_{ki} > b_{ki'}$ and $b_{k'i} < b_{k'i'}$. For the example above, these conditions are met for families 3357 and 3415 in both the linear and quadratic equations. In contrast, linear response curves do not intersect for families 3049 and 3415, but the quadratic curves do. Note that the interactions are not important from the standpoint of changes in rank; the curve intersections occur either at the extreme sites or outside the range of the data. However, it is potentially noteworthy that families that are significantly different

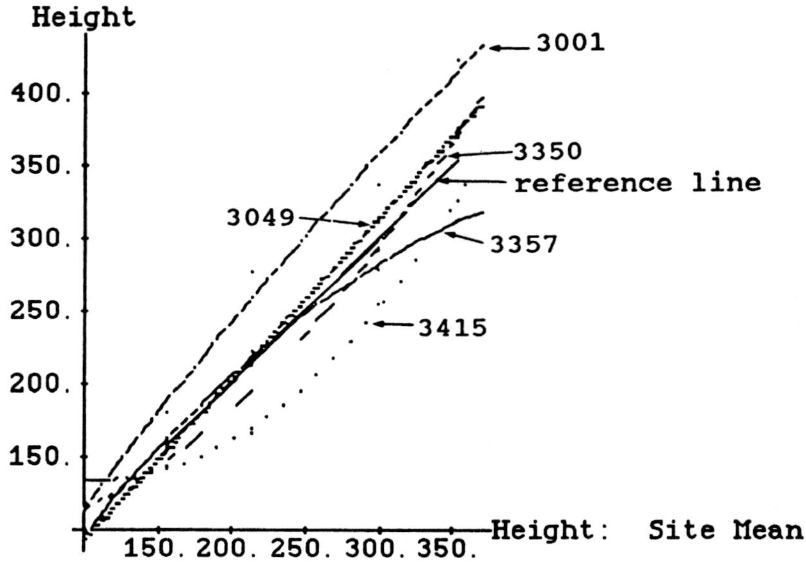

Figure 9.11. Quadratic functions of five half-sib families regressed against the plantation means.

at the intermediate sites might not be so at the best and poorest sites, and that there is some convergence of means at the best site.

9.3.3 Non-linear responses

Knight (1970), Gregorius and Namkoong (1986), and Namkoong et al. (1988) have argued that relative response curves are usually nonlinear and can take one of a number of forms against site means, reference genotypes, or direct measures of the environment. Knight (1973) suggested the use of polynomial equations to describe responses of parental genotypes versus their full-sib progeny over an array of environments. Linear, multiplicative, and exponential functions taken against a reference genotype have been proposed by Gregorius and Namkoong (1986).

The reference genotype circumvents the lack of independence in the Finlay/Wilkinson procedure as long as that genotype is included in all tests. To demonstrate the use of such a genotype, family 3350 from the example given in the previous section was specified as the reference; curves for quadratic regressions of this family against the remaining are shown in Fig. 9.12. The general form of these curves in relation to one another, as well as of those representing the environmental mean as reference, is similar to that for the exponential relationship in Gregorius and Namkoong (1986).

There are also alternatives to the arbitrarily chosen family illustrated above. Magnussen and Yeatman (1988) used a local seed source at each planting site as the reference. Thus a different genotypic entry represents the value of each planting site environment. Alternately, a check lot could be used as the reference. So, many genotypic entries and presumably, many sites can be compared as long as the same check lot is used in each breeding cycle.

Linear functions such as in the above example are probably adequate for selection in current breeding generations in the USA. However evaluation and selection programs in future generations will probably require a greater understanding of the nature of G x E interactions and consequently will have to rely on nonlinear models. Baker (1988)

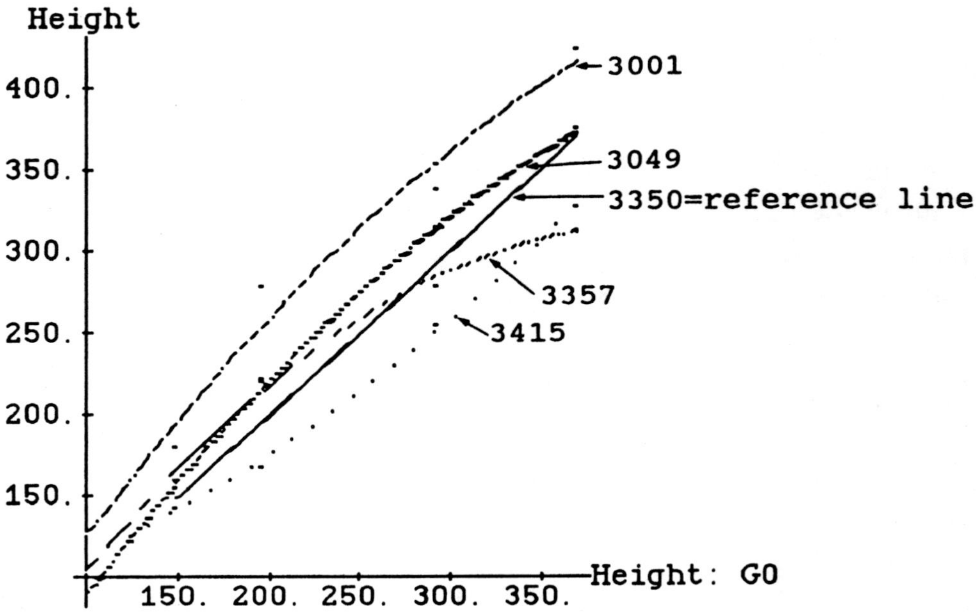

Figure 9.12. Quadratic functions of four half-sib families regressed against Family 3350 as reference genotype (G0).

discussed three such classes of these models: yield-density curves, as in growth and yield models (Ch 8); sigmoid curves, such as those describing growth; and asymptotic curves. I will discuss only the asymptotic function because it generally applies to fertilization responses, but has received little application in the evaluation of genotypes in tree breeding programs. The general form of the function is illustrated in Fig. 9.13, where the x axis defines some level of fertility and the y axis relative response.

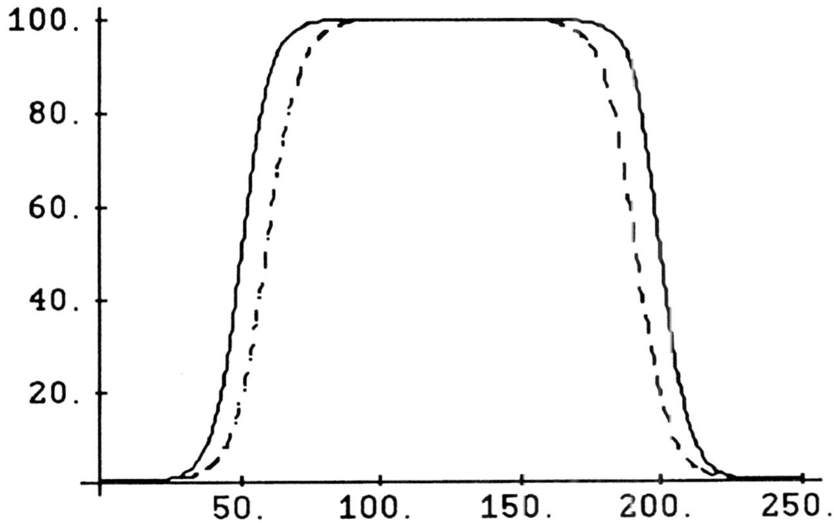

Figure 9.13. Asymptotic response curves for two hypothetical genotypes. For $0 < x < 125$, $y = 100/[1 + e^{10-cx}]$, and for $125 < x < 250$, $y = 100/[1 + e^{10-c(250-x)}]$. The solid curve is for rate constant c = 0.20 and the dashed curve for c = 0.17.

The left side of the curves is given by:

$$\frac{a}{1 + e^{(b - cx)}},$$

and the right side by:

$$\frac{a}{1 + e^{b-c(250-x)}},$$

where,

x = some specified environmental factor, such as levels of a nutrient.

Fig. 9.13 illustrates two genotypes that differ in rate constant. For one, c = 0.20 and for the other, c = 0.17. Both genotypes have the same maximum, but that defined by the solid line more efficiently responds to suboptimal levels of x than does the genotype defined by the dashed line. Moreover, the former genotype is more tolerant of superoptimal levels than the latter. Thus the former has a greater optimal range. Also important are intersections of the curves for genotypes, indicating interaction. This would occur if one genotype had a higher rate constant (c) and a lower asymptote (a) than another genotype.

To summarize, the linear methods and extensions are easiest to apply and superficially easier to interpret. However linearity of the response ought to be tested when more than five or so sites are assessed. If the assumption of linearity is found to be false, then one of quadratic approaches should be employed. And if comparisons among growth and yield or fertilizer responses are of interest, the appropriate nonlinear models should be used.

Focus in this section has been on parametric measures of stability. However, some nonparametric approaches have emerged in the literature over the last twenty years. One of these is to cluster genotypes on the basis of the similarity of response over site environments. This criterion is the converse of that described in

Abou-El-Fittouh et al. (1969), whereby clustering is on sites (Magnussen and Yeatman 1988). A discussion of alternative measures of similarity and distance is presented in Lin et al. (1986) along with a critical review of the parametric measures. Another approach is the biplot, whereby the first two principal components of the matrix of correlations among genotypic entries (say, families) over a range of sites are plotted (Corsten and Gabriel 1976; Gabriel 1971). Genotypic entries that perform similarly over sites will group together in the plot.

9.4 Future Directions and Needs

Most commonly-used experimental designs do not efficiently define the response of plants' genotypes to their environments. Because of demands for complex data, the same issues apply here as to those for testing seed sources through deployment of unbalanced designs (Friedman and Namkoong 1986, Namkoong, et al. 1988). For studies of environmental combinations, factorial designs are most suitable. However experimental sizes can exceed practical limits if more than two environmental factors are tested on more than a few genotypes. But fractional factorials (Box et al. 1978: Ch. 12) reduce the number of treatment combinations by confounding higher-order interactions with main effects and with lower-order interactions. For example, a $2^{(4-1)}$ fractional factorial would have two levels for each of four factors, but eight treatment combinations would be run instead of the usual 16. Thus in a fertilizer study, many combinations of nutrients can be tested simultaneously with minimal effort with many genotypes.

Series of factorials will then suggest which combinations of environmental variables contribute to responses in a set of traits. Then these variables would be entered as components of polynomial response surface models such as those in Section 9.2.2. However, biological interpretations are difficult to make from these mechanistic models. For this, nonlinear mechanistic models of G x E are required (Baker 1988; Box and Draper 1987; Box et al. 1978; Namkoong et al. 1988). For example, Foster (Ch. 7) advocated the use of growth and yield models to predict future stand development and resulting genetic gains from the deployment of particular sets of genotypes. There are further advantages of mechanistic models (Box et al. 1978: Ch. 16). They provide for extrapolation outside the data. And mechanistic models are usually simpler than empirical models, as witnessed by the large sizes of the polynomial equations in the examples in Sect. 9.2.2 versus three parameters in the Cauchy equation.

To build mechanistic models, screening studies designed to define important variables are first required. Then empirical models with these variables are developed. The form of these models should then lead to the formation of the mechanistic model.

The process of building a mechanistic model requires iterating the following steps: (1) parameter estimation by nonlinear regression, (2) model selection, and (3) tests of the stability of the model by examining its fit under extreme conditions. Experiments must be designed to discriminate and allow for choices among possible models. The choice of potential models will be guided by experiences from the other agricultural sciences. In other cases, models can be unique to woody plants.

In summary, a better understanding of processes such as competition and responses to fertilizers requires the use of mechanistic models. In using these, we will more accurately predict gains and risks. But the process of moving from standard quantitative genetic models demands changes both in experimental designs and strategies, and in statistical procedures.

Additional Reading

The following readings present introductory treatments of many of the subjects covered in this chapter.

Wright, J. W. 1978. *Introduction to Forest Genetics*. Academic Press, New York.

Zobel, B. J., and Talbert, J. 1984. *Applied Forest Tree Improvement*. J. Wiley & Sons, New York.

More detailed discussions of these subjects can be found in:

Namkoong, G. 1979. *Introduction to Quantitative Genetics in Forestry*. Tech. Bull. No. 1588, USDA Forest Service.

Namkoong, G., Kang, H. C., and Brouard, J. S. 1988. *Tree Breeding: Principles and Strategies*. Springer-Verlag, New York. See especially pages 103-115 and 132-149.

Thorough discussions of multivariate techniques are in:

Gittins, R. 1985. *Canonical Analysis. A Review with Applications in Ecology*. Springer-Verlag, New York.

Morrison, D. F. 1990. *Multivariate Statistical Methods*. McGraw-Hill, New York.

Response surface methodologies, with the focus primarily on fractional factorials and trend surface analysis are covered in:

Box, G. E. P., Hunter, W. G., and Hunter, J. S. 1978. *Statistics for Experimenters. An Introduction to Design, Data Analysis, and Model Building.* J. Wiley & Sons, New York.

For reviews of genotype by environment interactions and parametric measures of genotypic stability see:

Baker, R. J. 1988. Differential response to environmental stress. In *Proc. Second Int. Conf. Quant. Genetics* (Wier, B. S., Eisen, E. J., Goodman, M. M., and Namkoong, G., eds), pp. 492-504. Sinauer Assoc., Inc., Sunderland, MA.

Bridgwater, F. E., and Stonecypher, R. W. 1978. "Genotype x environment interaction: Implications in tree breeding programs." *Proc. Fifth N. Amer. For. Biol. Workshop*. pp. 46-63.

Matheson, A. C., and Cotterill, P. P. 1990. Utility of genotype-environment interactions. *For. Ecol. Manage.* 30:159-174.

Acknowledgements

I thank J. V. Brotschol, L. Fins, and S. T. Friedman for their patience and editorial assistance. Thanks are also due to R. K. Campbell, W. J. Libby, S. Magnussen, C. I. Millar, and G. Namkoong for many helpful comments. And I thank F. E. Bridgwater for alerting me to the biplot. M. T. Conkle generously provided the white fir range map.

Literature Cited

Abou-El-Fittouh, H. A., Rawlings, J. O., and Miller, P. A. 1969. Classification of environments to control genotype by environment interactions with an application to cotton. *Crop Sci.* 9:135-140.

Aitken, S. N., Westfall, R. D., and Libby, W. J. 1987. "Multivariate analysis of geographic variation in white fir *(Abies concolor)*." *Abstracts of Contributed Posters and Satellite Conferences, Second Int. Conf. Quant. Genetics*, Abstract 2, Raleigh, NC.

Baker, R. J. 1988. Differential response to environmental stress. In *Proc. Second Int. Conf. Quant. Genetics* (Wier, B. S., Eisen, E. J., Goodman, M. M., and Namkoong, G., eds), pp. 492-504. Sinauer Assoc., Inc., Sunderland, MA.

Box, G. E. P., and Draper, N. R. 1987. *Empirical Model-building and Response Surfaces*. J. Wiley & Sons, New York.

Box, G. E. P., Hunter, W. G., and Hunter, J. S. 1978. *Statistics for Experimenters. An Introduction to Design, Data Analysis, and Model Building*. J. Wiley & Sons, New York.

Burdon, R. D. 1977. Genetic correlation as a concept for studying genotype-environment interaction in forest tree breeding. *Silvae Genet*. 26(5-6):168-175.

Campbell, R. K. 1984. Procedures for determining the biological limits of breeding zones in the Pacific Northwest. In *Progeny Testing, Servicewide Genetics Workshop*, Charleston, SC. pp. 24-33.

Campbell, R. K. 1986. Mapped genetic variation of Douglas-fir to guide seed transfer in southwest Oregon. *Silvae Genet.* 35(2-3):85-96.

Conkle, M. T. 1973. Growth data for 29 years from the California elevational transect study of ponderosa pine. *For. Sci.* **19**:31-39.

Corsten, L. C. A., and Gabriel, K. R. 1976. Graphical exploration in comparing variance matrices. *Biometrics* **32**(6):851-863.

Cotterill, P. P. 1990. Short note: Numbers of families and progeny required for provenance testing. *Silvae Genet.* **39**(2):82-83.

Dorman, K. W. 1976. The Genetics and Breeding of Southern Pines. *Agric. Handbook 471*, USDA.

Draper, N. R., and Smith, H. 1966. *Applied Regression Analysis.* John Wiley & Sons, Inc., New York.

Eberhart, S. A., and Russell, W. A. 1966. Stability parameters for comparing varieties. *Crop Sci.* **6**:36-40.

Findlay, K. W., and Wilkinson, G. N. 1963. The analysis of adaptation in a plant breeding programme. *Aust. J. Agric. Res.* **14**:742-754.

Friedman, S. T., and Namkoong, G. 1986. Estimating family means using unbalanced incomplete blocks. In *IUFRO Conf., Joint Meeting of Working Parties on Breeding Theory, Progeny Testing, and Seed Orchards,* Williamsburg, VA. pp. 457-468.

Gabriel, K. R. 1971. The biplot graphic display of matrices with application to principal component analysis. *Biometrika* **58**(3):453-467.

Gittins, R. 1985. *Canonical Analysis. A Review with Applications in Ecology.* Springer-Verlag, New York.

Gregorius, H.-R., and Namkoong, G. 1986. Joint analysis of genotypic and environmental effects. *Theor. Appl. Genet.* **72**:413-422.

Harvey, W. R., and Townsend, A. M. 1985. Selection on a combination of individual, family, and stand merit in provenance tests. *Forest Sci.* **31**(4):813-821.

Kang, M. S. 1989. A new SAS program for calculating stability-variance parameters. *J. Hered.* **80**(5):415.

Kang, M. S., Miller, J. D., and Darrah, L. L. 1987. A note on relationship between stability variance and ecovalence. *J. Hered.* **78**:107.

Knight, R. 1970. The measurement and interpretation of genotype-environment interaction. *Euphytica* **19**:225-235.

Knight, R. 1973. The relation between hybrid vigour and genotype-environment interactions. *Theor. Appl. Genet.* **43**:311-318.

Kung, F. H., and Clausen, K. E. 1984. Graphic solution in relating seed sources and planting sites for white ash plantations. *Silvae Genet.* **33**(2-3):46-53.

Léon, J., and Becker, H. C. 1988. Repeatability of some statistical measures of phenotypic stability - correlations between single year results and multi years results. *Plant Breed.* **100**(2):137-142.

Lin, C. S., Binns, M. R., and Lefkovitch, L. P. 1986. Stability analysis: where do we stand? *Crop Sci.* **26**(5):894-900.

Magnussen, S., and Yeatman, C. W. 1988. Provenance hybrids in jack pine: 15-year results in eastern Canada. *Silvae Genet.* **37**(5-6):206-218.

Morrison, D. F. 1976. *Multivariate Statistical Methods.* McGraw-Hill Book Co., New York.

Namkoong, G. 1982. An application of biometrics in provenance analysis. *J. Tree Sci.* 1(1&2):57-63.

Namkoong, G. 1985. The influence of composite traits on genotype by environment relations. *Theor. Appl. Genet.* 70:315-317.

Namkoong, G., and Johnson, J. A. 1987. Influence of the value function on genotype-by-environment relations. *Silvae Genet.* 36(2):92-94.

Namkoong, G., Kang, H. C., and Brouard, J. S. 1988. *Tree Breeding: Principles and Strategies.* Springer-Verlag, New York.

Namkoong, G., Usanis, R. A., and Silen, R. R. 1972. Age-related variation in genetic control of height growth in Douglas-fir. *Theor. Appl. Genet.* 42:151-159.

Raymond, C. A., and Lindgren, D. 1986. A model for genetic flexibility. In *Provenances and Forest Tree Breeding of High Latitudes* (Lingren, D., ed), pp. 159-177. Swedish Univ. Agr. Sci., Dept. For. Genetics & Plant Physiol.,

Raymond, C. A., and Namkoong, G. 1990. Optimizing breeding zones: Genetic flexibility or maximum value. *Silvae Genet.* In press.

Rehfeldt, G. E. 1986. Adaptive variation in *Pinus ponderosa* from Intermountain regions. I. Snake and Salmon River basins. *Forest Sci.* 32(1):79-92.

Rehfeldt, G. E. 1988. Ecological genetics of *Pinus contorta* from the Rocky Mountains (USA): a synthesis. *Silvae Genet.* 37(3-4):131-135.

Rehfeldt, G. E. 1990. Gene resource management: Using models of genetic variation in silviculture. Genetics/Silviculture Workshop, Wenatchee, WA. In press.

Roberds, J., and Namkoong, G. 1986. Maximization of expected value for a trait in an environmental gradient. Ninth N. Amer. For. Biology Workshop, Stillwater, OK, June 15-18, 1986. pp. 179-186.

Roberds, J. H., and Namkoong, G. 1989. Population selection to maximize value in an environmental gradient. *Theor. Appl. Genet.* 77(1):128-134.

Rohlf, F. J., and Sokal, R. R. 1981. *Statistical Tables.* W. H. Freeman & Co., New York.

SAS Institute, I. 1985. *SAS® User's Guide: Statistics.* SAS Institute Inc, Cary, NC.

Shukla, G. K. 1972. Some statistical aspects of partitioning genotype-environmental components of variability. *Hered.* 29:237-245.

Steele, R. G. D., and Torrie, J. H. 1980. *Principles and Procedures of Statistics.* McGraw Hill Book Co., New York.

Thorpe, R. S. 1987. Complex clines: the predictivity of complicated patterns of geographic variation portrayed by multivariate analysis. *Biol. J. Linn. Soc.* 31:75-88.

Wartenberg, D. 1985. Canonical trend surface analysis: A method for describing geographic patterns. *Syst. Zool.* 34(3):259-279.

Westfall, R. D., Cody, J., and Hunt, C. M. 1977. Suggested tree planting zones and corresponding seed collection zones for the Northeast. In *A seed collection-seed bank and forest nursery system for the Northeast, Completion Report*, pp. 37-49. Applied Forestry Research Institute, SUNY College of Environmental Science and Forestry, Syracuse, NY.

Williams, B. K., and Titus, K. 1988. Assessment of sampling stability in ecological applications of discriminant analysis. *Ecology* 69(4):1275-1285.

Wolfram, S. 1988. *Mathematica™. A System for Doing Mathematics by Computer.* Addison-Wesley Publishing Company, Inc., Redwood City, CA.

Wricke, G. 1962. Über eine Methode zur Erfassung der ecologischen Streubreite in Feldversuchen. *A. Pflanzenzucht* 47:92-96.

Wright, J. W. 1978. *Introduction to Forest Genetics*. Academic Press, New York.

Appendix 9.1: Matrix Algebra Applied to Multivariate Analyses

Consider a 2 x 2 correlation matrix:

$$\mathbf{R} = \begin{bmatrix} 1 & \rho \\ \rho & 1 \end{bmatrix}.$$

Then eigenroots (or characteristic roots) of **R** are:

$$\lambda = [1 + \rho \quad 1 - \rho\ ].$$

And corresponding eigenvectors (or characteristic vectors) are:

$$\mathbf{V} = \begin{bmatrix} 1 & -\rho/(1 - (1 - \rho)) \\ 1 & -\rho/(1 - (1 + \rho)) \end{bmatrix}.$$

You can see that the second column in **V** reduces to -1 and +1.

A one standard deviation ellipsoid for $\rho = 0.50$ is shown below. The first eigenvector is aligned along the longest axis and the second is at right angles to the first. These are known as the principal components of this multivariate probability density distribution and vectors are known as principal vectors. Respective eigenroots represent the variance in each vector. The bivariate frequency distribution for this example is shown in the next figure.

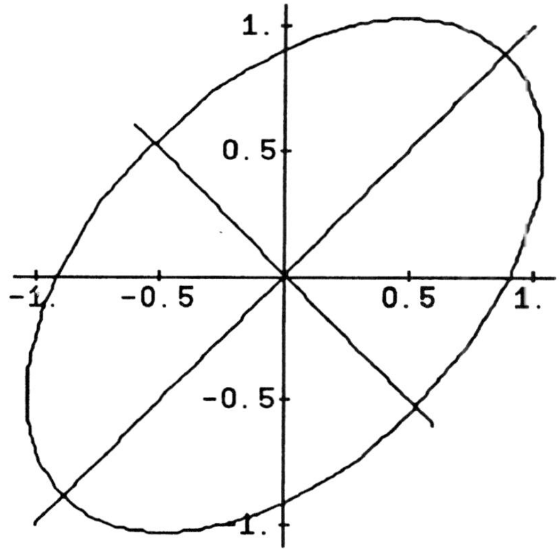

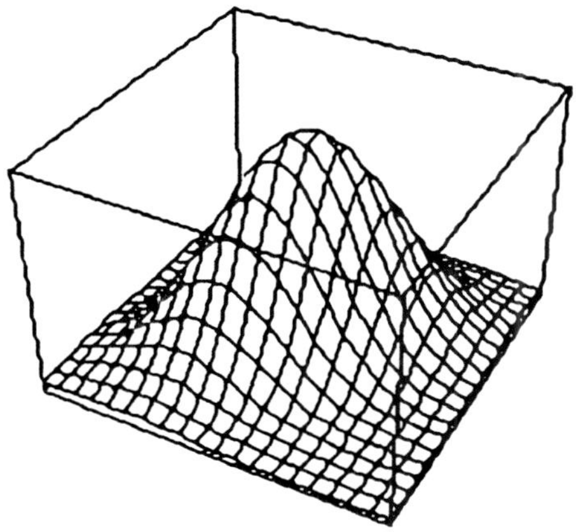

Now consider a three-way equicorrelation matrix:

$$\mathbf{R} = \begin{bmatrix} 1 & \rho & \rho \\ \rho & 1 & \rho \\ \rho & \rho & 1 \end{bmatrix}.$$

Its eigenvalues are:

$$\lambda = [1 + 2\rho \quad 1 - \rho \quad 1 - \rho\ ].$$

The eigenvectors for this matrix are more complex. However, the first vector, corresponding to the largest root is a vector of one's (1.0), as in the bivariate correlation matrix. In order to be more easily interpretable, the column vectors of **V** must be normalized or divided by the square root of the sum of squared column elements as in:

$$V_{xi} = \sqrt{\sum_{j=1}^{p} V_{ij}^2}.$$

Where p = the number of variables. Thus the transformation is:

$$V_n = VV_x^{-1}.$$

where $\mathbf{V}_n$ is the normalized eigenvectors and $\mathbf{V}_x^{-1}$ is the inverse of the diagonal matrix of the v_{xi}'s. The normalized first vector (transposed) for the preceding correlation matrix will thus be $[\ 1/\sqrt{3}\ \ 1/\sqrt{3}\ \ 1/\sqrt{3}]$. This illustrates the special properties of equicorrelation matrices, where the correlation coefficients are equal or approximately equal. For these matrices, the first root will be $[\ 1 + (p\ -\ 1)r]$ and $(\ 1\ -\ r)$ for the remaining roots. Also the normalized first eigenvector is:

$$v'_{n1} = \left[\frac{1}{\sqrt{p}}, \ldots, \frac{1}{\sqrt{p}}\right].$$

As a further example, let

$$\mathbf{R} = \begin{bmatrix} 1.000 & 0.974 & 0.973 \\ 0.974 & 1.000 & 0.965 \\ 0.973 & 0.865 & 1.000 \end{bmatrix}.$$

The diagonal matrix of eigenroots of **R** is:

$$\Lambda = \begin{bmatrix} 2.94 & 0.00 & 0.00 \\ 0.00 & 0.03 & 0.00 \\ 0.00 & 0.00 & 0.02 \end{bmatrix}.$$

The proportions of the total variance, described by the vectors are equal to $\Lambda/\mathrm{Tr}[\mathbf{R}]$ or 0.98, 0.01, and 0.01, respectively, where $\mathrm{Tr}[\mathbf{R}]$ is the trace of **R** or the sum of the diagonal elements of the matrix.

The eigenvectors of **R** are:

$$\mathbf{V} = \begin{bmatrix} 0.708 & 0.045 & 0.898 \\ 0.706 & 0.500 & -0.508 \\ 0.706 & -0.545 & -0.392 \end{bmatrix},$$

and

$$\mathbf{V}_n = \begin{bmatrix} 0.578 & 0.060 & 0.814 \\ 0.577 & 0.675 & -0.460 \\ 0.577 & -0.736 & -0.355 \end{bmatrix}.$$

This latter matrix is also known as the direction cosines of the vectors (you will see Namkoong frequently using such terminology). These are the cosines of the angles the vector takes in reference to the original vectors (or axes of the variables). This matrix is such that

$$\mathbf{V}_n'\mathbf{V}_n = \begin{bmatrix} 1.0 & 0.0 & 0.0 \\ 0.0 & 1.0 & 0.0 \\ 0.0 & 0.0 & 1.0 \end{bmatrix}.$$

That is, the vectors are orthogonal, or in other words, uncorrelated. Also,

$$\mathbf{V}_n^{-1}\Lambda\mathbf{V}_n = \mathbf{R}$$

Correlations between the original variables and the principal vectors are given by

$$\mathbf{V}_n\sqrt{\Lambda} = \begin{bmatrix} 0.992 & 0.011 & 0.126 \\ 0.990 & 0.126 & -0.071 \\ 0.989 & -0.137 & -0.055 \end{bmatrix}$$

These data are very informative. The first vector accounts for nearly all of the variation in the three variables, indicating that the ellipsoid describing the variables is nearly cigar-shaped: long, thin, and nearly round at the cross section at right angles to the longest axis.

The above concepts can also be applied to the products of variance-covariance or correlation matrices. For example, take a between-group variance-covariance matrix **B** and the pooled within-group matrix **W**. Ellipsoids for two groups are shown below. The first eigenvector of $\mathbf{BW}^{-1}$ will be aligned in the direction of the greatest difference between groups or as shown in the line angled from the lower left to the upper right in the figure. The second vector is at right angles to the first in this case.

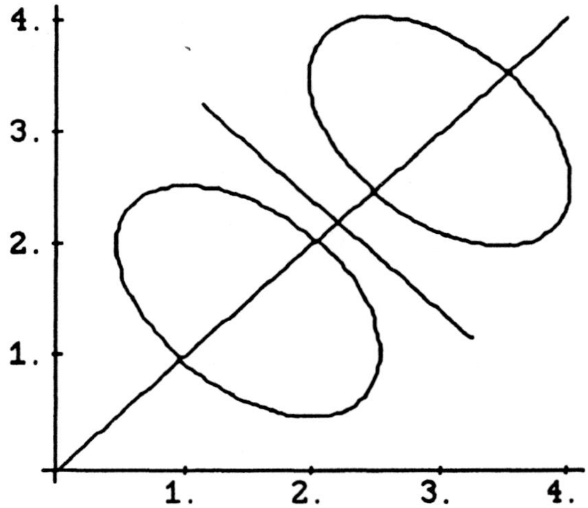

Appendix 9.2: Matrix Formulas for Multiple and Multivariate Regression

The linear regression model is (in matrix form):

$$\mathbf{y} = \mathbf{X}\beta + \varepsilon.$$

The vector of regression coefficients (β) is estimated by:

$$\beta = [X'\mathbf{X}]^{-1}\mathbf{X}'\mathbf{y}$$

where **X** is the n x k matrix of n observations of k dependent variables and **y** is the n x 1 vector of observations of the dependent variable.

The proportion of variance in **y** described by the model is:

$$\mathbf{R}^2 = (\beta'\mathbf{X}'\mathbf{y} - n\bar{\mathbf{y}}^2)/(\mathbf{y}'\mathbf{y} - n\bar{\mathbf{y}}^2).$$

Canonical correlation analysis:

This is the multivariate extension of multiple regression, above. With a matrix **X** of observations on p independent variables, and a corresponding matrix **Y** of q dependent variables, the partitioned matrix of correlations among these variables is:

$$\mathbf{R} = \begin{bmatrix} \mathbf{R}_{xx} & \mathbf{R}_{xy} \\ \mathbf{R}_{xy}^{T} & \mathbf{R}_{yy} \end{bmatrix}$$

where $\mathbf{R}^{T}$ is equal to $\mathbf{R}'$ or the transpose of the matrix.

Then let

$$\mathbf{F} = \mathbf{R}_{xx}^{-1} \mathbf{R}_{xy} \mathbf{R}_{yy}^{-1} \mathbf{R}_{xy}'$$

and the eigenvector decomposition is:

$$\mathbf{F} = \mathbf{V \Lambda V}^{-1}$$

where $\mathbf{\Lambda}$ is a diagonal matrix of the k eigenvalues of **F** (these are the squared canonical correlations), where k = min[p,q], and **V** is the column matrix of eigenvectors of **F**.

The matrices of canonical weights for **X** and **Y** are:

$$\mathbf{B} = \mathbf{R}_{xx}^{-0.5}\mathbf{V}$$

and

$$\mathbf{A} = \mathbf{R}_{yy}^{-1}\mathbf{R}'_{xy}\mathbf{B\Lambda}^{-0.5},$$

respectively. The matrices of structural correlations are:

$$\mathbf{U_y} = \mathbf{R_{yy}A}$$

and

$$\mathbf{U_x} = \mathbf{R_{xx}B}.$$

These are column vectors of correlations between the original variables and their own canonical vectors. These would be equivalent to the correlations between the independent variables and the predicted dependent variable, if you ever cared to compute them. Canonical scores are:

$$\mathbf{T_x} = \mathbf{X_sB} \qquad \text{Eq. 9.4}$$

and

$$\mathbf{T_y} = \mathbf{Y_sA} \qquad \text{Eq. 9.5}$$

where $\mathbf{X_s}$ and $\mathbf{Y_s}$ are the n x p and n x q standardized data matrices, respectively. Structure correlation matrices between the variables and the scores of the opposite canonical vectors are:

$$\mathbf{W_x} = \mathbf{U_x\Lambda^{0.5}}$$

and

$$\mathbf{W_y} = \mathbf{U_y\Lambda^{0.5}}$$

and are the multivariate equivalent of the vector of correlations between the dependent variable and the independent variables in multiple regression. The square of each correlation in **W** is the proportion of variance in the variable described by the model in each canonical vector. Thus the sum of these squared correlations over the vectors is the total proportion of variance described by the canonical analysis (referred to as the redundancies).

Subject Index

FORESTRY SCIENCES

1. P. Baas (ed.): *New Perspectives in Wood Anatomy*. Published on the Occasion of the 50th Anniversary of the International Association of Wood Anatomists. 1982
ISBN 90-247-2526-7
2. C.F.L. Prins (ed.): *Production, Marketing and Use of Finger-Jointed Sawnwood*. Proceedings of an International Seminar Organized by the Timber Committee of the UNECE (Halmar, Norway, 1980). 1982 ISBN 90-247-2569-0
3. R.A.A. Oldeman (ed.): *Tropical Hardwood Utilization*. Practice and Prospects. 1982
ISBN 90-247-2581-X
4. P. den Ouden (in collaboration with B.K. Boom): *Manual of Cultivated Conifers*. Hardy in the Cold- and Warm-Temperate Zone. 3rd ed., 1982
ISBN Hb 90-247-2148-2; Pb 90-247-2644-1
5. J.M. Bonga and D.J. Durzan (eds.): *Tissue Culture in Forestry*. 1982
ISBN 90-247-2660-3
6. T. Satoo: *Forest Biomass*. Rev. ed. by H.A.I. Madgwick. 1982 ISBN 90-247-2710-3
7. Tran Van Nao (ed.): *Forest Fire Prevention and Control*. Proceedings of an International Seminar Organized by the Timber Committee of the UNECE (Warsaw, Poland, 1981). 1982 ISBN 90-247-3050-3
8. J.J. Douglas: *A Re-Appraisal of Forestry Development in Developing Countries*. 1983
ISBN 90-247-2830-4
9. J.C. Gordon and C.T. Wheeler (eds.): *Biological Nitrogen Fixation in Forest Ecosystems*. Foundations and Applications. 1983 ISBN 90-247-2849-5
10. M. Németh: *Virus, Mycoplasma and Rickettsia Diseases of Fruit Trees*. Rev. (English) ed., 1986 ISBN 90-247-2868-1
11. M.L. Duryea and T.D. Landis (eds.): *Forest Nursery Manual*. Production of Bareroot Seedlings. 1984; 2nd printing 1987 ISBN Hb 90-247-2913-0; Pb 90-247-2914-9
12. F.C. Hummel: *Forest Policy*. A Contribution to Resource Development. 1984
ISBN 90-247-2883-5
13. P.D. Manion (ed.): *Scleroderris Canker of Conifers*. Proceedings of an International Symposium on Scleroderris Canker of Conifers (Syracuse, USA, 1983). 1984
ISBN 90-247-2912-2
14. M.L. Duryea and G.N. Brown (eds.): *Seedling Physiology and Reforestation Success*. Proceedings of the Physiology Working Group, Technical Session, Society of American Foresters National Convention (Portland, Oregon, USA, 1983). 1984
ISBN 90-247-2949-1
15. K.A.G. Staaf and N.A. Wiksten (eds.): *Tree Harvesting Techniques*. 1984
ISBN 90-247-2994-7
16. J.D. Boyd: *Biophysical Control of Microfibril Orientation in Plant Cell Walls*. Aquatic and Terrestrial Plants Including Trees. 1985 ISBN 90-247-3101-1
17. W.P.K. Findlay (ed.): *Preservation of Timber in the Tropics*. 1985
ISBN 90-247-3112-7
18. I. Samset: *Winch and Cable Systems*. 1985 ISBN 90-247-3205-0
19. R.A. Leary: *Interaction Theory in Forest Ecology and Management*. 1985
ISBN 90-247-3220-4
20. S.P. Gessel (ed.): *Forest Site and Productivity*. 1986 ISBN 90-247-3284-0

FORESTRY SCIENCES

21. T.C. Hennessey, P.M. Dougherty, S.V. Kossuth and J.D. Johnson (eds.): *Stress Physiology and Forest Productivity*. Proceedings of the Physiology Working Group, Technical Session, Society of American Foresters National Convention (Fort Collins, Colorado, USA, 1985). 1986 ISBN 90-247-3359-6
22. K.R. Shepherd: *Plantation Silviculture*. 1986 ISBN 90-247-3379-0
23. S. Sohlberg and V.E. Sokolov (eds.): *Practical Application of Remote Sensing in Forestry*. Proceedings of a Seminar on the Practical Application of Remote Sensing in Forestry (Jönköping, Sweden, 1985). 1986 ISBN 90-247-3392-8
24. J.M. Bonga and D.J. Durzan (eds.): *Cell and Tissue Culure in Forestry*. Volume 1: General Principles and Biotechnology. 1987 ISBN 90-247-3430-4
25. J.M. Bonga and D.J. Durzan (eds.): *Cell and Tissue Culure in Forestry*. Volume 2: Specific Principles and Methods: Growth and Development. 1987 ISBN 90-247-3431-2
26. J.M. Bonga and D.J. Durzan (eds.): *Cell and Tissue Culure in Forestry*. Volume 3: Case Histories: Gymnosperms, Angiosperms and Palms. 1987 ISBN 90-247-3432-0
    Set ISBN (Volumes 24-26) 90-247-3433-9
27. E.G. Richards (ed.): *Forestry and the Forest Industries: Past and Future*. Major Developments in the Forest and Forest Industries Sector Since 1947 in Europe, the USSR and North America. In Commemoration of the 40th Anniversary of the Timber Committee of the UNECE. 1987 ISBN 90-247-3592-0
28. S.V. Kossuth and S.D. Ross (eds.): *Hormonal Control of Tree Growth*. Proceedings of the Physiology Working Group, Technical Session, Society of American Foresters National Convention (Birmingham, Alabama, USA, 1986). 1987 ISBN 90-247-3621-8
29. U. Sundberg and C.R. Silversides: *Operational Efficiency in Forestry*. Vol. 1: Analysis. 1988 ISBN 90-247-3683-8
30. M.R. Ahuja (ed.): *Somatic Cell Genetics of Woody Plants*. Proceedings of the IUFRO Working Party S2.04-07 Somatic Cell Genetics (Grosshansdorf, Germany, 1987). 1988 . ISBN 90-247-3728-1
31. P.K.R. Nair (ed.): *Agroforestry Systems in the Tropics*. 1989 ISBN 90-247-3790-7
32. C.R. Silversides and U. Sundberg: *Operational Efficiency in Forestry*. Vol. 2: Practice. 1989 ISBN 0-7923-0063-7
    Set ISBN (Volumes 29 and 32) 90-247-3684-6
33. T.L. White and G.R. Hodge (eds.): *Predicting Breeding Values with Applications in Forest Tree Improvement*. 1989 ISBN 0-7923-0460-8
34. H.J. Welch: *The Conifer Manual*. Volume 1. 1991 ISBN 0-7923-0616-3
35. P.K.R. Nair, H.L. Gholz, M.L. Duryea (eds.): *Agroforestry Education and Training*. Present and Future. 1990 ISBN 0-7923-0864-6
36. M.L. Duryea and P.M. Dougherty (eds.): *Forest Regeneration Manual*. 1991 ISBN 0-7923-0960-X
37. J.J.A. Janssen: *Mechanical Properties of Bamboo*. 1991 ISBN 0-7923-1260-0
38. J.M. Bonga and P. Von Aderkas: *In Vitro Culture of Trees*. 1992 ISBN 0-7923-1540-5
39. L. Fins, S.T. Friedman and J.V. Brotschol (eds.): *Handbook of Quantitative Forest Genetics*. 1992 ISBN 0-7923-1568-5

KLUWER ACADEMIC PUBLISHERS – DORDRECHT / BOSTON / LONDON